住房和城乡建设部"十四五"规划教材
高等学校土木工程专业国际化人才培养全英文系列教材

Civil Engineering Drawing and Computer Drafting

土木工程制图与计算机绘图

常　虹　常鸿飞　主　编
卢丽敏　王　勇　任　宇　副主编
　　　　郑天翔　主　审

中国建筑工业出版社
China Architecture & Building Press

图书在版编目(CIP)数据

土木工程制图与计算机绘图 = Civil engineering drawing and computer drafting：英文/常虹，常鸿飞主编；卢丽敏，王勇，任宇副主编. — 北京：中国建筑工业出版社，2022.6

住房和城乡建设部"十四五"规划教材. 高等学校土木工程专业国际化人才培养全英文系列教材

ISBN 978-7-112-27269-3

Ⅰ.①土… Ⅱ.①常…②常…③卢…④王…⑤任… Ⅲ.①土木工程-建筑制图-高等学校-教材-英文②土木工程-计算机制图-高等学校-教材-英文 Ⅳ.①TU204

中国版本图书馆 CIP 数据核字(2022)第 056627 号

责任编辑：仕 帅 吉万旺
责任校对：姜小莲

住房和城乡建设部"十四五"规划教材
高等学校土木工程专业国际化人才培养全英文系列教材
Civil Engineering Drawing and Computer Drafting
土木工程制图与计算机绘图
常 虹 常鸿飞 主编
卢丽敏 王 勇 任 宇 副主编
郑天翔 主 审

*

中国建筑工业出版社出版、发行(北京海淀三里河路9号)
各地新华书店、建筑书店经销
北京科地亚盟排版公司制版
天津安泰印刷有限公司印刷

*

开本：787毫米×1092毫米 1/16 印张：16¾ 字数：488千字
2022年8月第一版 2022年8月第一次印刷
定价：58.00元 (赠教师课件)
ISBN 978-7-112-27269-3
(39153)

版权所有 翻印必究
如有印装质量问题，可寄本社图书出版中心退换
(邮政编码 100037)

Brief Introduction

This book is a textbook of engineering drawing courses for civil engineering majors in colleges and universities. It consists of 7 chapters. The main contents are: introduction, unified standards and basic skills for building drawings, architectural construction drawings, structural construction drawings, road route engineering drawings, bridge engineering drawings, computer aid drawings——AutoCAD, etc.

This book can be used as a teaching material for engineering drawing courses in civil engineering, Construceon management and other majors in colleges and universities. The book can also be referenced for students or engineering technicians of other related majors.

本书为高等院校土木类专业工程制图课程教材，共7章，主要内容有：绪论，房屋建筑制图统一标准及基本技能，建筑施工图，结构施工图，道路路线工程图，桥隧工程图以及计算机辅助制图等。

本书可作为高等院校土木工程、工程管理等专业的工程制图课程的教材，也可作为其他相关专业的学生或工程技术人员的参考书。

为了更好地支持教学，我社向采用本书作为教材的教师提供课件，有需要者可与出版社联系，索取方式如下：建工书院 http://edu.cabplink.com，邮箱 jckj@cabp.com.cn，电话（010）58337285。

出版说明

党和国家高度重视教材建设。2016年，中办国办印发了《关于加强和改进新形势下大中小学教材建设的意见》，提出要健全国家教材制度。2019年12月，教育部牵头制定了《普通高等学校教材管理办法》和《职业院校教材管理办法》，旨在全面加强党的领导，切实提高教材建设的科学化水平，打造精品教材。住房和城乡建设部历来重视土建类学科专业教材建设，从"九五"开始组织部级规划教材立项工作，经过近30年的不断建设，规划教材提升了住房和城乡建设行业教材质量和认可度，出版了一系列精品教材，有效促进了行业部门引导专业教育，推动了行业高质量发展。

为进一步加强高等教育、职业教育住房和城乡建设领域学科专业教材建设工作，提高住房和城乡建设行业人才培养质量，2020年12月，住房和城乡建设部办公厅印发《关于申报高等教育职业教育住房和城乡建设领域学科专业"十四五"规划教材的通知》（建办人函〔2020〕656号），开展了住房和城乡建设部"十四五"规划教材选题的申报工作。经过专家评审和部人事司审核，512项选题列入住房和城乡建设领域学科专业"十四五"规划教材（简称规划教材）。2021年9月，住房和城乡建设部印发了《高等教育职业教育住房和城乡建设领域学科专业"十四五"规划教材选题的通知》（建人函〔2021〕36号）。为做好"十四五"规划教材的编写、审核、出版等工作，《通知》要求：（1）规划教材的编著者应依据《住房和城乡建设领域学科专业"十四五"规划教材申请书》（简称《申请书》）中的立项目标、申报依据、工作安排及进度，按时编写出高质量的教材；（2）规划教材编著者所在单位应履行《申请书》中的学校保证计划实施的主要条件，支持编著者按计划完成书稿编写工作；（3）高等学校土建类专业课程教材与教学资源专家委员会、全国住房和城乡建设职业教育教学指导委员会、住房和城乡建设部中等职业教育专业指导委员会应做好规划教材的指导、协调和审稿等工作，保证编写质量；（4）规划教材出版单位应积极配合，做好编辑、出版、发行等工作；（5）规划教材封面和书脊应标注"住房和城乡建设部'十四五'规划教材"字样和统一标识；（6）规划教材应在"十四五"期间完成出版，逾期不能完成的，不再作为《住房和城乡建设领域学科专业"十四五"规划教材》。

住房和城乡建设领域学科专业"十四五"规划教材的特点：一是重点以修订教育部、住房和城乡建设部"十二五""十三五"规划教材为主；二是严格按照专业标准规范要求编写，体现新发展理念；三是系列教材具有明显特点，满足不同层次和类型的学校专业教学要求；四是配备了数字资源，适应现代化教学的要求。规划教材的出版凝聚了作者、主审及编辑的心血，得到了有关院校、出版单位的大力支持，教材建设管理过程有严格保障。希望广大院校及各专业师生在选用、使用过程中，对规划教材的编写、出版质量进行反馈，以促进规划教材建设质量不断提高。

住房和城乡建设部"十四五"规划教材办公室
2021年11月

Preface

The Belt and Road (B&R), is a global infrastructure development initiative adopted by our government in 2013. A 2019 study conducted by global economic consultants forecasted that the B&R was likely to boost world GDP by ＄7.1 trillion per annum by 2040. Examples of B&R infrastructure investments include ports, skyscrapers, railroads, roads, bridges, airports, dams, coal-fired power stations, and railroad tunnels. To meet the needs of the overseas construction market for professionals who master professional English, this textbook is specially compiled.

This textbook is based on the basic requirements for the teaching of engineering graphics courses in general colleges and universities and the requirements for the basic quality of engineering and technical personnel in the field of engineering formulated by the higher education engineering graphics teaching guiding committee of the Ministry of Education. It is written on the basis of the textbook *Civil Engineering Drawing and Computer Drafting* (Chinese version) funded by the New Century Textbook Construction Project of China University of Mining and Technology.

This textbook covers architectural engineering, road engineering, bridge and tunnel engineering and AutoCAD in content. The content is to be concise and practical, the structure is reasonable, and the teaching content is aimed at meeting the actual needs of engineering applications. This textbook refers to the latest current drawing standards and collections, including *Unified Standard for Building Drawings* (GB/T 50001—2017), *Standard for General Layout Drawings* (GB/T 50103—2010), *Standard for Architectural Drawings* (GB/T 50104—2010), *Standard for Structural Drawings* (GB/T 50105—2010), *Standard for Road Engineering Drawings* (GB 50162—1992) and *Drawing Rules and Standard Detailing Drawings of Ichnographic Representing Method for Construction Drawings of R.C. Structures* (16G101).

This textbook is edited by Chang Hong of China University of Mining and Technology. Authors participating in the compilation work are Chang Hong and Ding Wei (Chapter 1, Chapter 2), Chang Hong (Chapter 4, Sections 4.1~4.6), Chang Hongfei (Chapter 3), Wang Yong (Chapter 4, Section 4.7, Chapter 5), Lu Limin (Chapter 6), Lu Limin and Ren Yu from CUMT Engineering Consultancy & Research Institute Co., LTD (Jiang Su) (Chapter 7). Thanks to Zheng Tianxiang, an expert from Poly International Holdings Co., Ltd for reviewing this textbook.

During the writing and publication of this textbook, the authors have fortunately received different kinds of help and support from their colleagues and peers, and they would like to express heartfelt thanks here. Due to the limited level of authors, omissions and deficiencies in the textbook are inevitable, and any valuable comments and suggestions from the readers are appreciated.

Authors
2022.3

前言

"一带一路"是我国政府在2013年制定的一项全球基础设施发展倡议。2019年的一项研究预测，到2040年，"一带一路"可能会使世界GDP每年增加7.1万亿美元。"一带一路"基础设施投资包括：港口、大楼、铁路、公路、桥梁、机场、大坝、燃煤发电站和铁路隧道。为适应"一带一路"政策下海外建筑市场对掌握专业英语人才的需求，特编写本教材。

本书是根据教育部工程图学教学指导委员会制定的普通高等院校工程图学课程教学基本要求以及工程领域对工程技术人员基本素质的要求，在中国矿业大学新世纪教材建设工程资助教材《土木工程制图与计算机绘图》的基础上编写而成。

本书在内容上涵盖了建筑工程、道路工程、桥隧工程等专业，内容力求简洁、实用，结构编排合理，教学内容以满足工程应用实际需求为目的。本书参考了现行最新的制图规范及图集，包括：《房屋建筑制图统一标准》GB/T 50001—2017、《总图制图标准》GB/T 50103—2010、《建筑制图标准》GB/T 50104—2010、《建筑结构制图标准》GB/T 50105—2010、《道路工程制图标准》GB 50162—1992以及《混凝土结构施工图平面整体表示方法制图规则和构造详图》16G101系列图集。

本书由中国矿业大学常虹主编。参与编写工作的有中国矿业大学常虹和丁威（第1章、第2章）、常鸿飞（第3章）、常虹（第4章4.1~4.6节）、王勇（第4章4.7节、第5章）、卢丽敏（第6章）、卢丽敏和任宇（中国矿业大学工程咨询研究院（江苏）有限公司）（第7章）。感谢保利国际控股有限公司专家郑天翔对本书的审阅。

在本书的编写和出版过程中，作者有幸得到了同事和同行的各种帮助和支持，在此表示衷心的感谢。由于作者水平有限，本教材难免有遗漏和不足之处，敬请广大读者提出宝贵意见和建议。

作　者
2022年3月

Contents

Chapter 1 Introduction
1.1 Properties and Tasks of the Course ... 2
1.2 Contents and Requirements of the Course ... 2
1.3 Projection Methods ... 3
1.4 Classification of Civil Engineering Drawings ... 4
1.5 Development of Construction Engineering Drawings in China ... 7

Chapter 2 Unified Standards and Basic Skills for Building Drawings
2.1 United Standard for Drawings ... 10
2.2 Common Drawing Instruments, Tools, and Their Usages ... 25
2.3 Drawing Steps and Methods ... 28

Chapter 3 Architectural Construction Drawings
3.1 Overview ... 30
3.2 Regulations of Architectural Construction Drawings ... 35
3.3 Drawing List and General Construction Instructions ... 47
3.4 Construction Site Plan ... 50
3.5 Floor Plan ... 54
3.6 Elevation Drawings ... 67
3.7 Architectural Section ... 73
3.8 Architectural Details ... 78

Chapter 4 Structural Construction Drawings
4.1 Overview ... 86
4.2 General Requirements in Standard for Structural Drawings of Buildings ... 93
4.3 General Notes of Structure Design ... 98
4.4 Ichnographic Construction Drawings of Foundations ... 99
4.5 Ichnographic Construction Drawings of the Main Structure ... 125
4.6 Ichnographic Construction Drawings of Stairs ... 142
4.7 Construction Drawings of Steel Structures ... 150

Chapter 5 Road Route Engineering Drawings
5.1 Highway Route Drawings ... 165
5.2 Urban Road Route Drawings ... 174

Chapter 6 Bridge Engineering Drawings
6.1 Bridge Engineering Drawings ... 180
6.2 Tunnel Engineering Drawings ... 188

Chapter 7　Computer Aid Drawings——AutoCAD

　7.1　Introducing AutoCAD　　196
　7.2　Introducing Drawings　　201
　7.3　Draw tools, Object Snap and Dynamic Input　　211
　7.4　Zoom, Pan and Templates　　215
　7.5　The Modify Tools　　222
　7.6　Dimensions and Text　　237
　7.7　Hatching　　247
　7.8　Building Drawings　　253

References

Chapter 1
Introduction

1.1 Properties and Tasks of the Course

In civil engineering, it is necessary to accurately describe the size, shape, layout, materials, and other properties of the object to be constructed for the design and construction of buildings (residential buildings and public buildings), industrial structures (plants and warehouses), roads, bridges or tunnels, etc. Even the most detailed descriptions or texts can not make people fully understand a project to be constructed. The information of a project can not be accurately expressed with general texts. However, engineering drawings can clearly show the information and construction requirements of a project, so they are used as the construction basis. Engineering drawings, which are also called "the technical language in engineering", are drawn by uniform provisions of the national, regional, and industrial standards. They are the critical basis for engineering construction or manufacturing.

All the technicians of civil engineering must be good at drawing and reading engineering drawings. They should be able to express design concepts through plotting drawings and understand others' design ideas through reading drawings. Therefore, Civil Engineering Drawing is one of the fundamental courses in the teaching plan of civil engineering.

This course details the theories and methods of reading and plotting engineering drawings and aims to cultivate students' basic abilities of reading and plotting civil engineering drawings. The course can lay students a good foundation for advanced courses. The main tasks of the students in the course are introduced as follows:

1) To learn national drawing standards and other relevant provisions.

2) To properly use drawing instruments and tools and master drawing skills.

3) To master the legends and requirements of professional engineering drawings, including architectural construction drawings, structural construction drawings, highway engineering drawings, and bridge and tunnel drawings.

4) To develop the basic skills of drawing and reading the specialized drawings.

5) To master the skills in computer graphics.

1.2 Contents and Requirements of the Course

This course includes three parts. The first part introduces common standards and basic skills of building drawings. The second part, the core part of this course, details various specialized drawings including architectural construction drawings, structural construction drawings, highway engineering drawings, and bridge and tunnel drawings. The third part introduces computer graphics. The contents and requirements of each part are as follows:

1) Through learning common standards and basic skills for building drawings, students are required to be familiar with national standards of engineering drawings (such as the provision of the sheet, title bar, signature column, lines, scale, fonts, and dimension mark), properly use drawing instruments and tools and mater basic skills of plotting drawings.

2) Through learning specialized drawings, students are required to master relevant standards of specialized drawings, and graphical presentation methods, contents and characteristics of the architectural construction drawings, structural construction drawings, highway engi-

neering drawings, bridge, and tunnel drawings and be familiar with the basic steps and methods of reading and plotting engineering drawings.

3) Through learning computer graphics, students are required to master basic operations of CAD (including drawing commands, editing commands, text input, dimension marking, and printing).

1.3 Projection Methods

A projection method, as the basis of descriptive geometry, is the method of mapping a three-dimensional object to a two-dimensional plane. The projection is obtained from shadows through a series of assumptions and abstraction, such as the shadow of a single object in the sun or the light. According to different positions of the projection center relative to the projection plane, projection methods may be classified into prospective (central) projection methods and parallel projection methods. The latter may be further classified into the orthographic projection method and the oblique projection method.

1. Perspective (Central) projection

When the distance from the center of projection (COP) to the projection plane (PP) is finite and the projection lines are converged on a point, the projection method is called the perspective projection method or central projection method. The drawing plotted according to this projection method is called the perspective drawing.

As shown in Figure 1.3.1, the projection of $\triangle ABC$, $\triangle A'B'C'$, is formed through the projection from Point S onto plane P. Point S is called COP and Plane P is called PP. Lines SA, SB, and SC are called projection lines. In short, $\triangle A'B'C'$ is the projection of $\triangle ABC$ on Plane P. In Figure 1.3.1, all projection lines converge at the center of projection, Point S, so this method is called the perspective projection method or central projection method. The obtained projection is also called the central projection.

2. Parallel projection

When the distance from COP to PP is infinite and the projection lines are parallel to each other, the projection method is called the parallel projection method, as shown in Figure 1.3.2. According to different projection angles, parallel projections may be classified into the oblique projection method and the orthographic projection method.

When the projection line intersects PP with an angle that is not equal to 90° (Figure 1.3.2a), this parallel projection method is called the oblique projection method. The drawing obtained

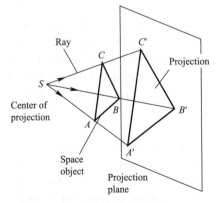

Figure 1.3.1 Perspective projection

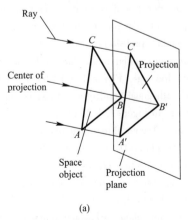

Figure 1.3.2 Parallel projection (one)
(a) Oblique projection method;

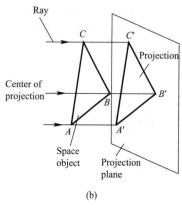

Figure 1.3.2 Parallel projection (two)
(b) Orthographic projection method

by the oblique projection method is called the oblique projection drawing. This projection method is generally used for axonometric projection.

When the projection line and the PP form a right angle (Figure 1.3.2b), this parallel projection method is called the orthographic projection method, and the corresponding drawing is called the orthographic projection drawing. The orthographic projection method can reflect the real shape and size of an object. It is a simple drawing method. Therefore, it is the main method for various engineering drawings.

1.4 Classification of Civil Engineering Drawings

In civil engineering, it is necessary to draw various drawings of buildings, such as houses, roads, bridges, culverts, and tunnels. Different drawings have different characteristics and requirements, so different graphical presentation methods are required. The common projection drawings include multi-plane orthogonal projection drawings, perspective projection drawings, axonometric projection drawings, and elevation projection drawings.

1. Multi-plane orthogonal projection drawings

The multi-plane orthogonal projection method is to project an object onto at least two mutually perpendicular projection planes by using the orthographic projection method. The obtained drawings are called multi-plane orthogonal projection drawings. Figure 1.4.1 shows the four-plane orthogonal projection drawings of one building. In Figure 1.4.2(a), a cup foundation is projected onto three mutually perpendicular projection planes H, V, and W. After expanding the three projection planes, the three-plane orthogonal projection drawings are obtained, as shown in Figure 1.4.2(b).

In multi-plane orthogonal projection drawings, each view can only represent the dimensions of the object in two directions. Therefore, it lacks the cubic effect. However, it can accurately and completely show the surface shape and size of an object. It is simple for drawing and measurement. Therefore, the multi-plane orthogonal projection drawing is one of the most important drawings in civil engineering and will be highlighted in the book.

2. Perspective projection drawings

Perspective projection is to project an object onto a projection plane (PP) with the central projection. The obtained drawing is called the perspective projection drawing or perspective drawing, as shown in Figure 1.4.3.

The projection drawing obtained when the relative position of COP, the object, and PP are properly set is consistent with the observed common image. The projection drawing is characterized by the cubic effect. However, it can not be measured or reflect the real shape and size of an object. It is a difficult drawing method. Figure 1.4.4 shows a perspective projection drawing of a building.

During the architectural design, perspective projection drawings are generally used as the basis to investigate the building's shape and further adjust and modify the design scheme. Therefore, perspective projection drawings are often used as schematic drawings to display

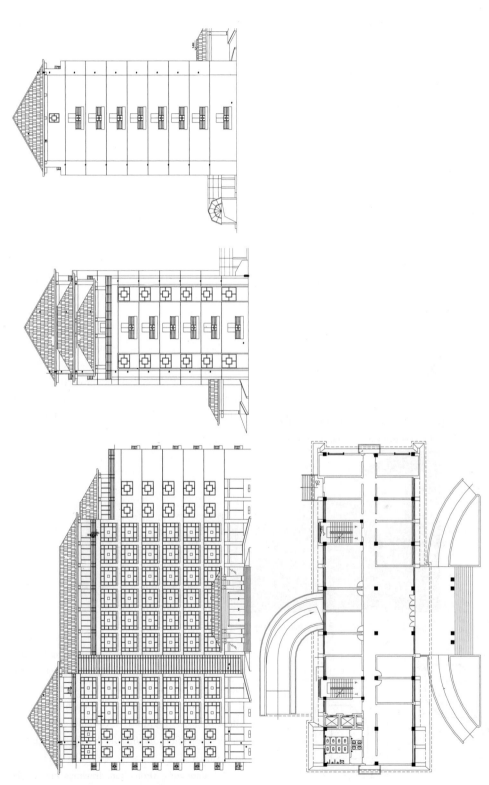

Figure 1.4.1　Four-plane orthogonal projection drawing of a building

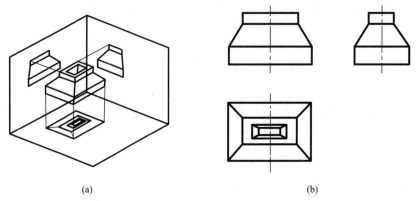

Figure 1.4.2 Three-phase orthogonal projection drawing of a cup foundation
(a) An object orthogonally projected onto three planes; (b) Orthogonal projection drawings by expanding projection planes

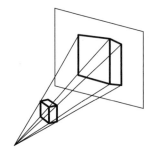

Figure 1.4.3 Form of a projection drawing

buildings' 3D renderings. Perspective projection is one of the essential aided means during architectural design.

3. Axonometric projection drawings

Axonometric projection is to project an object and its axes onto a PP by using the parallel projection method, where the axes of the object are not parallel to the PP. The obtained drawing is called the axonometric projection drawing or the axonometric drawing. As shown in Figure 1.4.5, the PP is called the projection plane. The axes O_1X_1, O_1Y_1, and O_1Z_1 are projected onto the plane P, and OX, OY, and OZ are called axonometric axes.

Figure 1.4.4 Perspective projection drawing of buildings

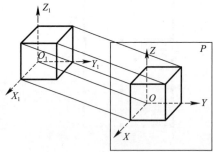

Figure 1.4.5 Formation of an axonometric projection drawing

The axonometric projection drawing has the cubic effect. Although the effect is not better than that of perspective projection drawings, the drawing method is easier. The measurement effect is poor. Figure 1.4.6 shows the axonometric projection drawing of the cup foundation in Figure 1.4.2.

Axonometric projection drawings are auxiliary drawings in civil engineering and are usually used to draw piping drawings, such as water supply and drainage, heating and ventilation, air conditioning drawings.

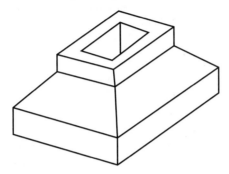

Figure 1.4.6 An axonometric projection drawing of a cup foundation

4. Elevation projection drawings

As an orthographic projection method, elevation projection is to project contour lines of the ground onto a horizontal projection plane, where height values are given to display topographical conditions. The obtained drawings are called elevation projection drawings. As shown in Figure 1.4.7, the sparseness of contour lines may be used to evaluate the topographical conditions of an area.

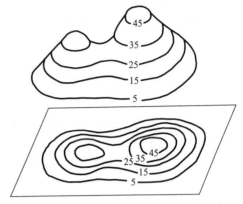

Figure 1.4.7 Elevation projection drawing

In civil engineering, elevation projection drawings are usually used to show the topographic maps, plane layouts of roads and hydraulic engineering, etc.

1.5 Development of Construction Engineering Drawings in China

The technology of construction engineering drawings in China has been developing since 1950. In 1956, to satisfy the economic construction demands of socialism, National Construction Committee approved the Monochrome Architectural Drawing Standards and Construction and Engineering Design Administration released Engineering Drawing Interim Standards. On this basis, the State Infrastructure Commission approved and promulgated *Standard for Architectural Drawing* (GBJ 1—73). Since 1986, the general parts have been modified and enhanced. State Planning Commission approved and implemented *United Standard for Building Drawings* (GBJ 1—86) together with professional drawing standards, including *Standard for General Layout Drawing* (GBJ 103—87), *Standard for Architectural Drawing* (GBJ 104—87), *Standard for Structural Drawings* (GBJ 105—87), *Standard for Building Water Supply and Drainage Drawings* (GBJ 106—87) and *Standard for Heating, Ventilation and Air Conditioning Drawings* (GBJ 114—88).

In 2001, the above-mentioned six drawing standards were modified and approved by the Ministry of Construction and other related departments. *United Standard for Building Drawings* (GB/T 50001—2001) was approved. *Standard for General Layout Drawings* (GB/T 50103—2001), *Standard for Architectural Drawings* (GB/T 50104—2001), *Standard for Structural Drawings* (GB/T 50105—2001), *Standard for Building Water Supply and Drainage Drawings* (GB/T 50106—2001), and *Standard for Heating, Ventilation and Air Conditioning Drawings* (GB/T 50114—2001) have been approved and implemented since March 1, 2002.

According to the requirements of Notice of 2007 Engineering Construction Standards Specification and Modification (First batch) issued by the Ministry of Construction, ***Unified Standard for Building Drawings*** (GB/T 50001—2010) was approved by China Building Standard Design & Research Institute (CBS) and other related units based on a series of standards approved in 2001. ***Standard for General Layout Drawings*** (GB/T 50103—2010), ***Standard for Architectural Drawings*** (GB/T 50104—2010), ***Standard for Structural Drawings*** (GB/T 50105—2010), ***Standard for Building Water Supply and Drainage Drawings*** (GB/T 50106—2010), and ***Standard for Heating, Ventilation and Air Conditioning Drawings*** (GB/T 50114—2010) have been approved and implemented since March 1, 2011. ***Unified Standard for Building Drawings*** (GB/T 50001—2017) has been approved and implemented since May 1, 2018.

With the popularization and extensive applications of computers, computer graphics has also developed rapidly. Computer graphics and CAD (computer-aided design) have been widely applied in civil engineering. The automation of drawing technology will be realized in construction engineering to satisfy the requirements of modernized construction.

Chapter 2
Unified Standards and Basic Skills for Building Drawings

Engineering drawings are used to express engineering design intentions. They are the important technical documents in construction, production, and management. They include not only the drawings obtained by the projection principle, but also the sizes, materials, methods, text descriptions of a project.

To realize the consistent understanding of technical drawing among technicians at different posts and allow technicians to express technical ideas through drawings, it is necessary to develop unified regulations of engineering drawings, which are the basis of drawing or reading drawings. These unified regulations are drawing standards.

The standards are generally formulated by designated national agencies, so they are called "national standards" and the code name is "GB". There are various national standards. To distinguish different technical standards, some letters and numbers are added after the code name. For example, the general code name of building standards is "GBJ".

The current national drawing standards in construction engineering have six volumes, including *Unified Standard for Building Drawings* (GB/T 50001—2017), *Standard for General Layout Drawings* (GB/T 50103—2010), *Standard for Building Drawings* (GB/T 50104—2010), *Standard for Architectural Drawings* (GB/T 50105—2010), *Standard for Building Water Supply and Drainage Drawings* (GB/T 50106—2010) and *Standard for HVAV Drawings* (GB/T 50114—2010).

Civil engineering drawings involve building engineering, water conservancy engineering, and road and bridge engineering. Therefore, this book introduces different drawing standards in corresponding chapters. This chapter mainly introduces *United Standard for Building Drawings* (GB/T 50001—2017) and briefly introduces common drawing instruments, methods, and steps. Students are expected to master drawing methods and skills. Unless otherwise stated, the standard in this chapter refers to *United Standard for Building Drawings* (GB/T 50001—2017).

2.1 United Standard for Drawings

1. Drawing sheet

To use drawing sheets reasonably and bind drawings easily, the drawing sheet is stipulated in *Unified Standard for Building Drawings* GB/T 50001—2017. A drawing sheet refers to the size of drawing paper. The drawing frame refers to the boundary of the drawing range. The basic sizes of drawing paper include A0, A1, A2, A3, and A4. The sizes of the drawing sheet and drawing frame are stipulated in *Unified Standard for Building Drawings* GB/T 50001—2017, as shown in Table 2.1.1 and Figure 2.1.1.

Drawing sheet and frame size (mm)　　　　Table 2.1.1

Size code \ Drawing sheet	A0	A1	A2	A3	A4
$b \times l$	841×1189	594×841	420×594	297×420	210×297
c	10			5	
a	25				

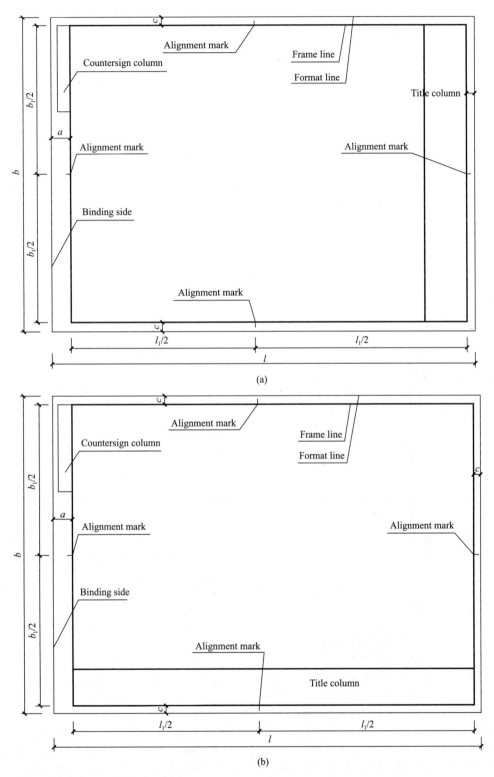

Figure 2.1.1 Formats of drawings (one)
(a) A0-A3 drawing of the horizontal type (1) (b) A0-A3 drawing of the horizontal type (2)

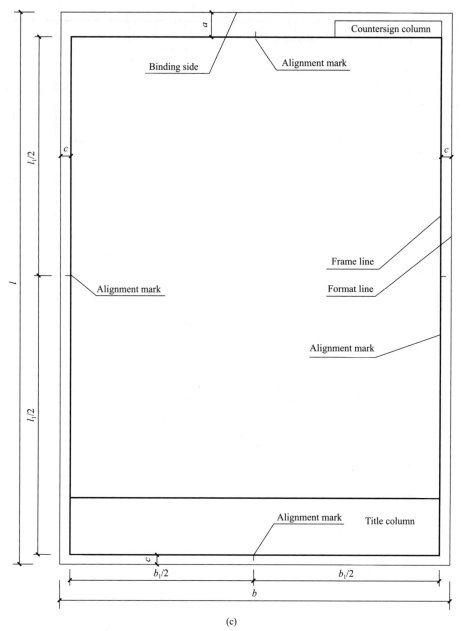

Figure 2.1.1 Formats of drawings (two)
(c) A0-A4 drawing of the vertical type (1)

The drawing sheet with short sides as vertical edges is called the horizontal type. The drawing sheet with short sides as horizontal edges is called the vertical type. In general, drawing paper A0–A3 should be used as the horizontal type. If necessary, it can also be used as the vertical type.

The side length of the drawing sheet should meet $l=\sqrt{2}\,b$. The area of A0 is $1\,m^2$. The area of A1 is half of A0, and so on.

If necessary, long sides of the drawing sheet may be lengthened, but short sides should not be extended. The size of the drawing sheet after extending long sides should meet the regulations in Table 2.1.2.

Drawing sheets should be chosen based on

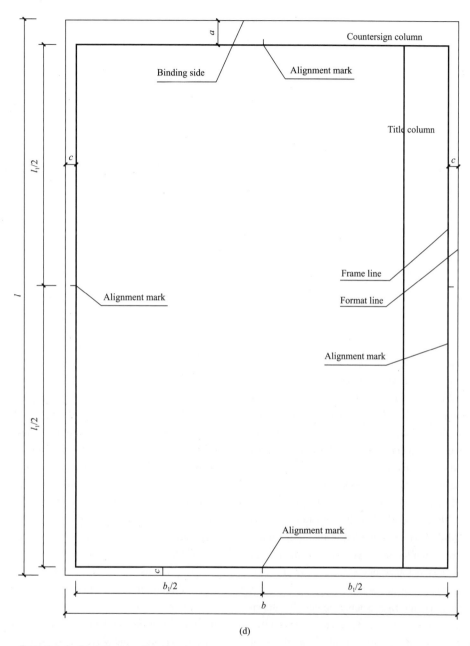

Figure 2.1.1 Formats of drawings (three)
(d) A0-A4 drawing in vertical type (2)

the complexity of drawings and should be united for convenience in construction sites. In engineering design, in addition to A4 for the contents and tables, the number of drawing sheets for each specialty should be no more than two.

2. Title column and countersign column

Each drawing includes the title column, frame lines, drawing lines, binding edges, and alignment mark. Each formal engineering drawing should include the project name, the design unit, drawing title, number, and signatures of designer, mapper, checker, reviewer, etc., which may be found in the right corner of the drawing, which is called the title column. The

Sizes of the drawing with the extended long side (mm) Table 2.1.2

Drawing sheet	Long side length	Length of the extended long side
A0	1189	1486(A0+1/4) 1635(A0+3/8) 1783(A0+1/2) 1932(A0+5/8) 2080(A0+3/4) 2230(A0+7/8) 2378(A0+1)
A1	841	1051(A1+1/4) 1261(A1+1/2) 1471(A1+3/4) 1682(A1+1) 1892(A1+5/4) 2102(A1+3/2)
A2	594	743(A2+1/4) 891(A2+1/2) 1041(A2+3/4) 1189(A2+1) 1338(A2+5/4) 1486(A2+3/2) 1635(A2+7/4) 1783(A2+2) 1932(A2+9/4) 2080(A2+5/2)
A3	420	630(A3+1/2) 841(A3+1) 1051(A3+3/2) 1261(A3+2) 1471(A3+5/2) 1682(A3+3) 1892(A3+7/2)

P.S.: For special drawings, $b \times l$ of 841mm×891mm and 1189mm×1261mm may be used.

countersign column is a table where each project director signs their names, date, etc. The positions of the title column and the countersign column are shown in Figures 2.1.1(a) and (b) for the horizontal type and Figures 2.1.1(c) and (d) for the vertical type.

The title column should be drawn according to Figure 2.1.2. The size, forms, and position of the title column are determined according to the requirements of a project. The signature area should contain the real name column and signature column. The title column of a foreign project should contain the translation below the Chinese. The "People's Republic of China" should be added above or on the left of the design unit.

In computer drawings, the electronic signature should follow the regulations of the Law of Electronic Signatures.

The countersign column should be drawn according to Figure 2.1.3, where the specialty, name, and date (year, month, and day) of participants are included. If one countersign column is not enough, another one may be increased. The countersign column is optional.

The title column and countersign column may be set according to the actual conditions of corresponding departments. The formats and contents of the title column may be designed by yourself in the homework of drawing and it is not necessary to set the countersign column in the homework of drawing.

3. Drawing lines

1) Line width and types

To show different contents clearly, different types of lines should be used in engineering drawings.

The line width in drawings should be 1.4, 1.0, 0.7, 0.5, 0.35, 0.25, 0.18 or 0.13mm. The line width should not be less than 0.1mm. In each drawing, the basic line width b should be firstly chosen according to the complexity and scale of the drawing, and then the corresponding line width set may be chosen from Table 2.1.3. In the same drawing, the drawings with the same scale should adopt the same line width set.

The line widths in Table 2.1.4 are applicable to frame lines and title column lines.

There are many line types in engineering drawings, such as solid line, dash line, chain line, break line, and wave lines. We can use different line types according to different conditions, as shown in Table 2.1.5.

2) Drawing methods of lines

Notes for drawing lines are provided as follows:

(1) The net space among parallel legend lines should be no less than 0.2mm.

(2) The lengths and intervals of dotted lines, long-dash dot lines, and long-dash double-dot lines should be the same.

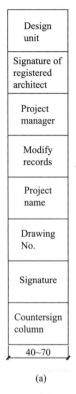

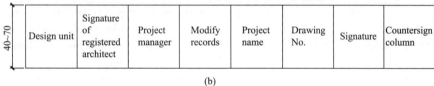

Figure 2.1.2　Title column
(a) vertical title column; (b) horizontal title column

Specialty	Name	Signature	Date

Figure 2.1.3　Countersign column

(3) Solid lines can replace long-dash dot lines and long-dash double-dot lines, when it is difficult to draw those lines in the smaller drawing.

(4) The ends of the long-dash dot line and long-dash double-dot line should not be a dot.

(5) The intersection point between dash-dot lines or the intersection point between a dash-dot line and other drawing lines should be formed with line segments.

(6) The intersection point between the dash lines or the intersection point between the dashed line and other drawing lines should be

formed with line segments.

(7) When a dashed line is the extension part of a solid line, it can not be connected with a solid line.

Line width set (mm)　　Table 2.1.3

Line width ratio	Width set			
b	1.4	1.0	0.7	0.5
$0.7b$	1.0	0.7	0.5	0.35
$0.5b$	0.7	0.5	0.35	0.25
$0.25b$	0.35	0.25	0.18	0.13

P. S.: 1. The miniature drawing should not adopt the line width of less than 0.18mm;

2. In the same drawing, the same thin line in the thinner line width set may be used as the thin line in different line width sets.

Widths of frame line and title column lines
Table 2.1.4

Drawing sheet code	Frame line	Outer frame line of title column	Sub-grid lines in the title column
A0, A1	b	$0.5b$	$0.25b$
A2, A3, and A4	b	$0.7b$	$0.35b$

(8) When drawing the centerline of a circle or an arc, the center of the circle should be the intersection point of lines.

(9) Lines should not overlap the characters, numbers, or symbols in the drawing. The characters must be clear. Table 2.1.6 shows some proper examples.

Drawing lines　　Table 2.1.5

Names		Types	Widths	General purposes
Solid line	Bold	———————	b	The main visible contour line
	Semi-bold	———————	$0.7b$	The visible contour line
	Middle	———————	$0.5b$	The visible contour line, dimension line, and change cloud line
	Thin	———————	$0.25b$	Legend filling line, furniture line
Dash line	Bold	- - - - - - - -	b	See the drawing standards of the related specialty
	Semi-bold	- - - - - - - -	$0.7b$	Invisible contour line
	Middle	- - - - - - - -	$0.5b$	Invisible contour line, legend line
	Thin	- - - - - - - -	$0.25b$	Legend filling line, furniture line
Long-dash dot line	Bold	—·—·—·—	b	See the drawing standards of the related specialty
	Middle	—·—·—·—	$0.5b$	See the drawing standards of the related specialty
	Thin	—·—·—·—	$0.25b$	The centerline, symmetric line, and axes, etc.

Continued

Names		Types	Widths	General purposes
Long-dash double-dot line	Bold	 —‥—‥—‥—	b	See the drawing standards of the related specialty
	Middle	 —‥—‥—‥—	$0.5b$	See the drawing standards of the related specialty
	Thin	 —‥—‥—‥—	$0.25b$	The supposed contour line, the original contour line
Break line		 ⟋	$0.25b$	Boundary line
Wave line		 ～	$0.25b$	Boundary line

Drawing methods of lines Table 2.1.6

Lines	Wrong	Right	Statement
Dash-dot line			The length and interval of the dotted line, long-dash dot line, and long-dash double-dot line should be the same. The ends of the long-dash dot line and long-dash double-dot line should not be a dot
Connection of lines	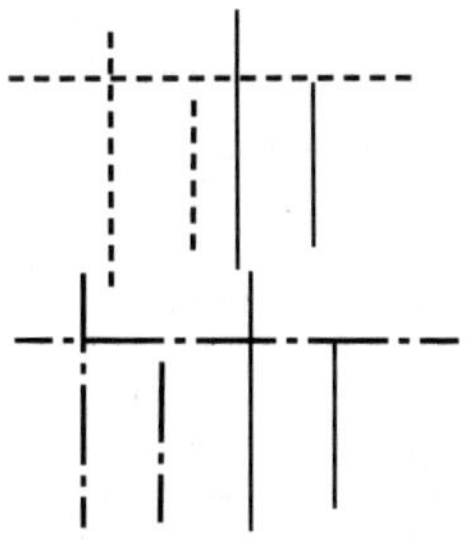	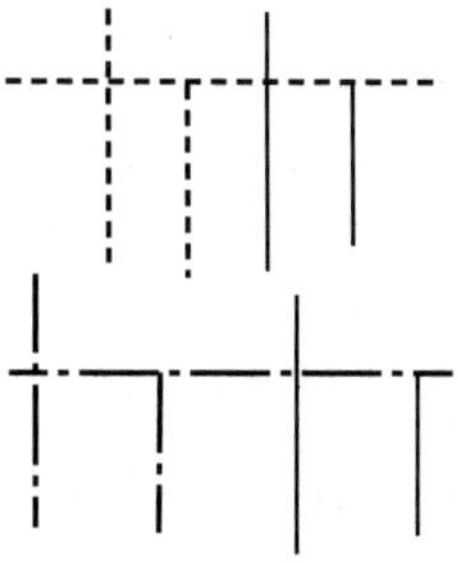	The intersection point between the dash-dot lines (dash lines) or between the dash-dot line (dash lines) and another drawing line should be the intersection point of lines
The centerline of a circle	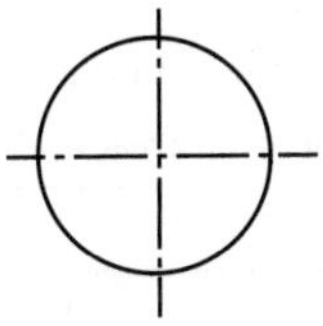	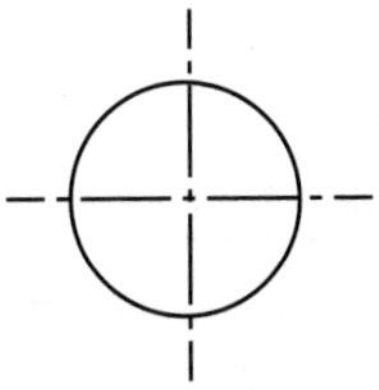	When drawing the centerline of a circle or an arc, the center of the circle should be the intersection point of lines
Break line and wave line	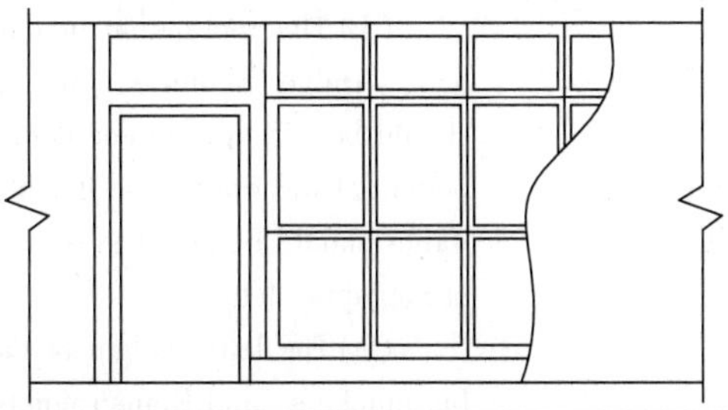		The symbols among the break line and wave lines are drawn by hand. The break line should be through the whole figure and exceed 2–3mm on both sides. The wave line shall not exceed the contour line

4. Fonts

The texts in engineering drawings usually include Chinese characters, Arabic numbers, Latin letters, Greek letters, etc. The writing should be clear. The punctuation should be used properly.

The text height should be chosen from Table 2.1.7. The text height of more than 10mm should use Truetype font. If a larger text is required, the height should be increased by time $\sqrt{2}$.

Height of texts (mm)
Table 2.1.7

Font type	Chinese vector font	Truetype font and non-Chinese vector font
Text height	3.5, 5, 7, 10, 14, 20	3, 4, 6, 8, 10, 14, 20

1) Chinese characters

Chinese characters in drawings should be written in the form of Fangsongti (vector font) or bold font. In the same drawing, the number of fonts should be no more than two. Table 2.1.8 shows the relations between the width and height of Fangsongti. However, the width and height of the bold font should be the same. Other legible fonts should be adopted in the headline, cover of the drawing album, and topographic map. The writing of simplified Chinese characters should meet related national regulations.

Relationship between the height and width of Fangsongti (mm)
Table 2.1.8

Heights	20	14	10	7	5	3.5
Widths	14	10	7	5	3.5	2.5

The grid should be drawn first before writing Fangsongti. The basic requirements of strokes include straight horizontal and vertical lines, symmetrical structure, and stuffed grids. Finally, the grids should be wiped off, as shown in Figure 2.1.4.

2) Numbers and letters

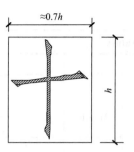

Figure 2.1.4 Basic requirements of Fangsongti

In engineering drawings, Latin letters, Arabic numbers, and Roman numbers should be written in the form of a simplified style or Roman font. Their rules of writing should conform to regulations in Table 2.1.9.

Writing rules of Latin letter, Arabic numbers, and Roman numbers
Table 2.1.9

Writing forms	General fonts	Narrow fonts
The height of the capital letter	h	h
The height of lowercase (no extension from top to bottom)	$7/10h$	$10/14h$
Extended head or trailing portion of lowercase	$3/10h$	$4/14h$
Stroke width	$1/10h$	$1/14h$
Letter space	$2/10h$	$2/14h$
The minimum space between datum lines in the downstream and upstream	$15/10h$	$21/14h$
Word space	$6/10h$	$6/14h$

3) Notes

(1) The inclination of italics of Latin letters, Arabic numbers, and Roman numbers should be 75° upward counterclockwise from the bottom of the word. The height and width of the italics should be equal to those of corresponding orthographic letters.

(2) The letter height of Latin letters, Arabic numbers, and Roman numbers should be no

less than 2.5mm.

(3) A number should be written in the form of Arabic numbers. All kinds of measurement units with a certain number should be written in the form of Unit symbols issued by the government.

(4) Grade, percentage, and scale numbers should be written in the form of Arabic numbers and mathematical symbols. For example, three fourths, twenty-five percent, and one to twenty should be written in the forms of 3/4, 25%, and 1 : 20.

(5) When a number is less than 1, "0" should be written out. The decimal point should be written in the form of a dot.

(6) Long Fangsongti characters, Latin letters, Arabic numbers, and Roman numbers should satisfy the requirements of **National Standards of Technical Drawing Font** (GB/T 14691—1993).

4) Application scopes of various font sizes (Table 2.1.10)

Application scopes of various font sizes
Table 2.1.10

Font size	Application
2.5, 3.5 size	Dimension mark and elevation
3.5, 5 size	Drawing name, Scale, General text description, Component number, Index symbol, Profile symbol, etc.
5, 7 size	Table title, Detail-drawing symbol, etc.
7, 10 size	Titles of various drawings
10, 14 size	Headline or cover title

The writing examples of letters and numbers with different font sizes are shown in Figure 2.1.5.

Figure 2.1.5 Writing examples of letters and numbers

5. Scale

In civil engineering drawings, it is necessary to scale down engineering drawings on the drawing. However, small engineering accessories are often required to be enlarged on the drawing. The linear-dimension ratio of a drawing corresponding to an actual object is called the scale of the drawing. For example, a drawing of 1 : 50 is larger than that of 1 : 100 and a drawing of 1 : 1 is the original drawing. The ratio larger than 1 is called the enlarging scale, such as 2 : 1. The ratio less than 1 is called the scaled-down scale, such as the ratio of 1 : 2.

The scale used in the drawing should be preferentially selected from Table 2.1.11 according to the purposes and complexity. Other scales can also be selected in special cases when the scales are given in a drawing.

In general, one scale is adopted in one drawing. According to drawing requirements, two kinds of scales may be adopted in the same drawing.

Scales in drawings

Table 2.1.11

Common scales	1∶1, 1∶2, 1∶5, 1∶10, 1∶20, 1∶30, 1∶50, 1∶100, 1∶150, 1∶200, 1∶500, 1∶1000, 1∶2000
Available scales	1∶3, 1∶4, 1∶6, 1∶15, 1∶25, 1∶40, 1∶60, 1∶80, 1∶250, 1∶300, 1∶400, 1∶600, 1∶5000, 1∶10000, 1∶20000, 1∶50000, 1∶100000, 1∶200000

The scale should be written on the right of the drawing name. The font height of the scale should one or two font sizes smaller than that of the drawing title. A solid line should be drawn under the drawing name as shown in Figure 2.1.6.

When only one scale is used in a drawing, the scale should be written in the title column.

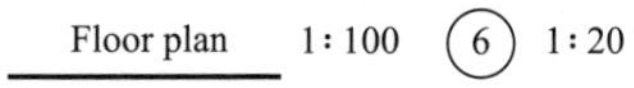

Floor plan   1∶100   ⑥   1∶20

Figure 2.1.6   Writing of scale

6. Dimension mark

The components in engineering drawings include extension lines, dimension lines, start-stop symbols, and dimension figures, as shown in Figure 2.1.7.

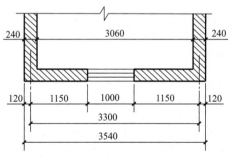

Figure 2.1.7   Sizes of different parts in a drawing

The requirements of each component are as follows:

1) Dimension lines

Dimension lines should be drawn in the form of a thin solid line. Dimension lines should be parallel to the contour line. The lines in drawings can not be used as dimension lines.

2) Extension line

Extension lines should be drawn in the form of a solid thin line. Extension lines should be perpendicular to dimension lines and beyond dimension lines about 2–3mm. There is about 2mm away from contour lines, as shown in Figure 2.1.8. Contour lines may be used as extension lines.

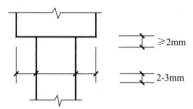

Figure 2.1.8   Boundary lines

3) Start-stop symbol

The start-stop symbol should be drawn in the form of a semi-bold solid line with a length of 2–3mm. The inclined direction of start-stop symbols should be clockwise 45° with boundary lines. Start-stop symbols of radii, diameters, angles, and arcs should be expressed with arrows, as shown in Figure 2.1.9.

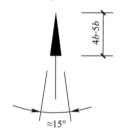

Figure 2.1.9   Start-stop symbols

4) Dimension figures

The size of drawings should be based on dimension figures, which are independent of the drawing scale. The size of drawings should not be directly measured from drawings. In construction drawings, the unit of millimeter must be adopted for all the components except the elevation and general layout. The unit should be provided in drawings.

The directions of dimension lines include horizontal, vertical, inclined directions, and dimension figures are written in the way shown in Figure 2.1.10(a). When the angle of inclination from the vertical direction is less than 30°, dimension figures should be written from the left side. If the angle of inclination from the vertical direction is in the sectional angle of 30°, dimension figures should be written in the way shown in Figure 2.1.10(b).

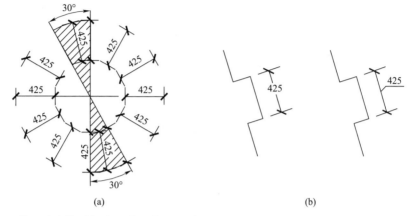

Figure 2.1.10 Directions of the dimension figures

Dimension figures are generally written in the upper middle of the dimension line. If there is not enough space, they may be written outside of the boundary line or written above and below the dimension line and adjacent dimension figures in the middle should be written above the outgoing line in a staggered way, as shown in Figure 2.1.11. They should be written from the left to the right above the horizontal dimension line. Dimension figures on the vertical dimension line should be written from the bottom to the top.

Figure 2.1.11 Positions of dimension figures

5) Arrangement of dimension lines

Dimension figures should be marked outside of the contour line. They should not intersect with lines, texts, and symbols. When such an intersection is inevitable, the lines should be interrupted to ensure the legibility of texts (Figure 2.1.12).

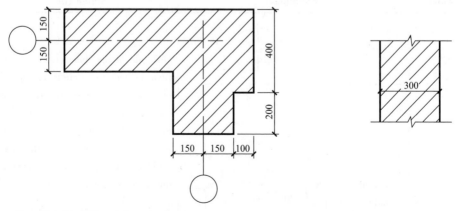

Figure 2.1.12 Writing of dimension figures

Parallel dimension lines should be arranged in the direction away from contour lines. The smaller sizes should be close to the contour line, as shown in Figure 2.1.13.

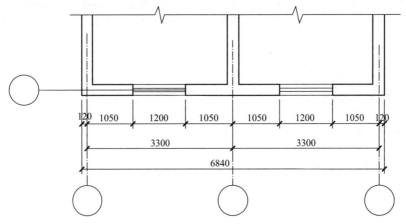

Figure 2.1.13　Arrangement of dimension lines

The distance between dimension lines and the contour lines should be no less than 10mm. The spacing between parallel dimension lines should be 7–10mm, and the same spacing value should be adopted in a drawing.

The boundary line of the total dimension should be close to contour lines. Boundary lines in the middle may be shorter, but the length should be the same (Figure 2.1.13).

6) Dimensions of radii, diameters, angles, arc length, and chord length

Table 2.1.12 shows the examples of dimension figures of the radius, diameter, angle, arc length, and chord length.

When the radius of a circle is marked, "SR" should be added in front of dimension figures. For marking the diameter, "$S\phi$" should be added in front of dimension figures. The marking methods are the same as the dimension marking methods of the radius and diameter of a circle or arc.

7) Other regulations of dimension mark

In addition to the above basic requirements, other requirements of dimension marking methods have been proposed in some standards (Table 2.1.13).

Dimension marking methods of radii, diameters, angles, arc length, and chord length

Table 2.1.12

Name	Examples		Statement
Radius	Radius		One side of the dimension line starts from the center of a circle and the other endpoints to the arc by arrows. "R" should be added in front of the number. As for larger arcs, the broken line or disconnected radius lines could be used. For small arcs, the radius may be marked outside
	Great-circle radius		
	Small arc radius		

Continued

Name		Examples	Statement
Diameter	Diameter	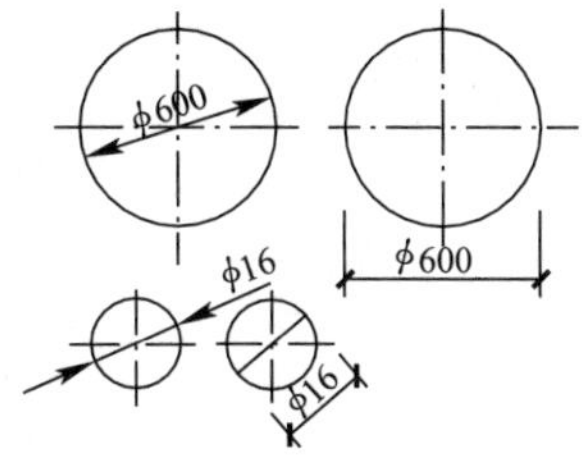	When the diameter of a circle is marked, the "ϕ" should be added in front of the number. The dimension line should pass the center of the circle with two ends pointing to the arc with an arrow. As for smaller diameters, the diameter may be marked outside
	Diameter of a small circle	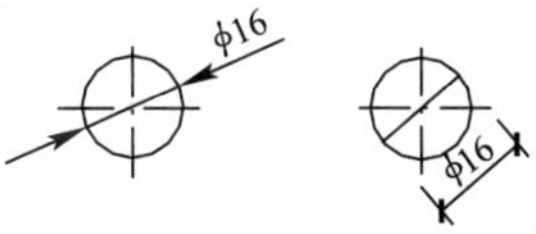	
Angle	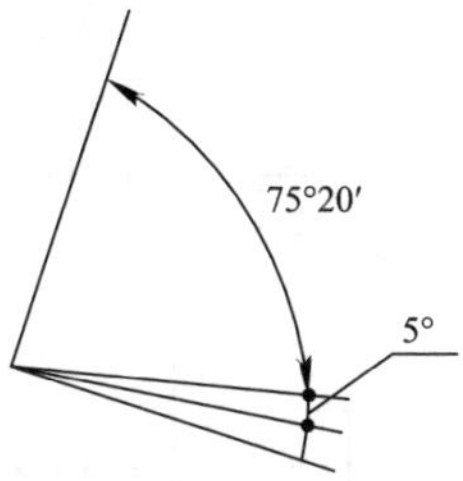		The dimension line of the angle is the arc. The center of the arc is the vertex of the angle. The arrows are used as the start-stop symbol. If there is not enough space, arrows may be replaced by dots. The angle should be marked in the top right corner
Arc length	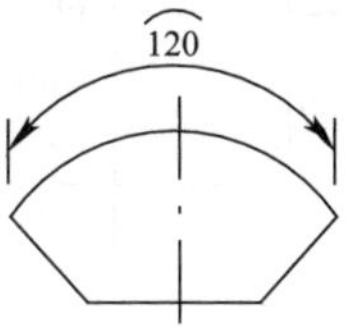		When the arc length is marked, the dimension line shall be marked by the arc. The boundary line should be perpendicular to the chord of the arc. The arrows are used as the start-stop symbols. "⌒" should be added on top of the dimension figure
Chord length	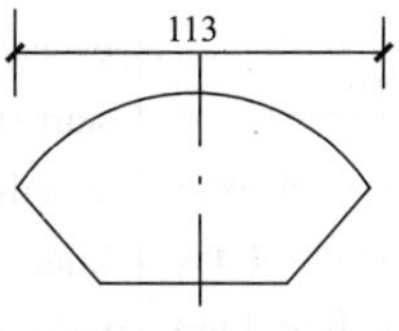		The dimension line should be a line parallel to the chord. The boundary line should be perpendicular to the chord. The bold virgule is used to show the start-stop symbols

Other requirements of dimension marking methods in some standards

Table 2. 1. 13

Contents	Examples	Making methods
The thickness of a thin plate	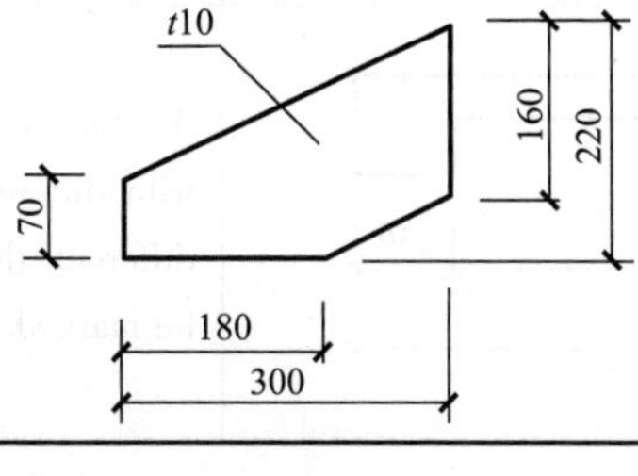	When the thickness of a thin plate is marked, "t" shall be added in front of the dimension figure

Continued

Contents	Examples	Making methods
Slope size	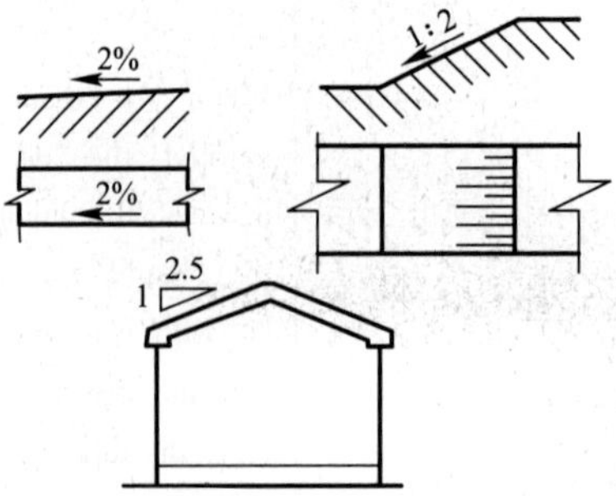	The single arrow is used as the slope symbol. The arrow should point to the downhill direction with the dimension figure on the arrow side or at the end. The slope can also be marked with the right triangle
Single line diagram	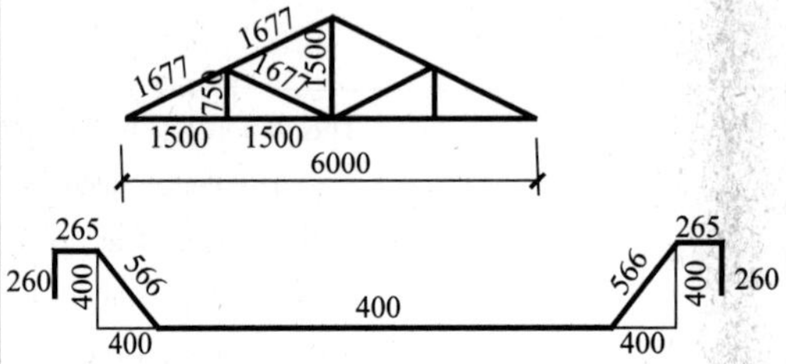	When the length of a rod or pipeline is marked, the figure may be marked along the rod
Isometric size	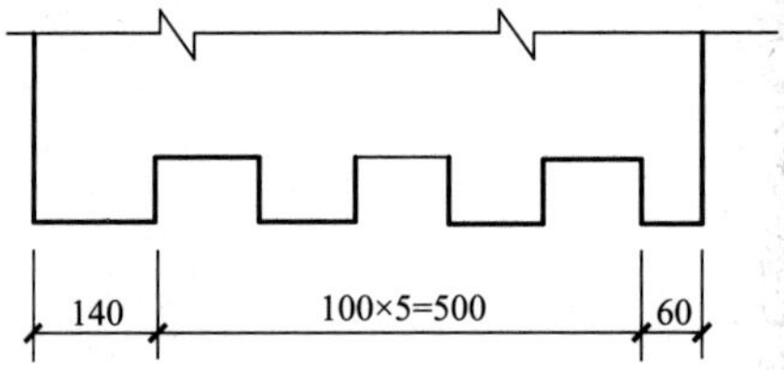	"Size × number = total length" may be used to indicate a series of consecutive components of the same size
Symmetrical components	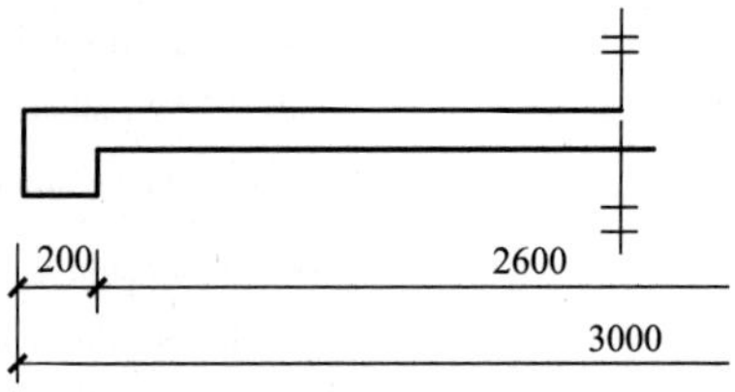 P. S. : Symmetrical symbols are composed of symmetric lines and two pairs of parallel lines at the end. The parallel lines are drawn with fine lines with a length of 6–10mm. The space should be 2–3mm. The symmetric lines vertically bisect the two parallel lines and are 2–3mm beyond the parallel lines at the end	The dimension lines of symmetrical components should slightly exceed symmetrical symbols. The start-stop symbol should be marked at only one side of the dimension line. The full-size figures should be written and their positions should be aligned with symmetric symbols
Two similar components	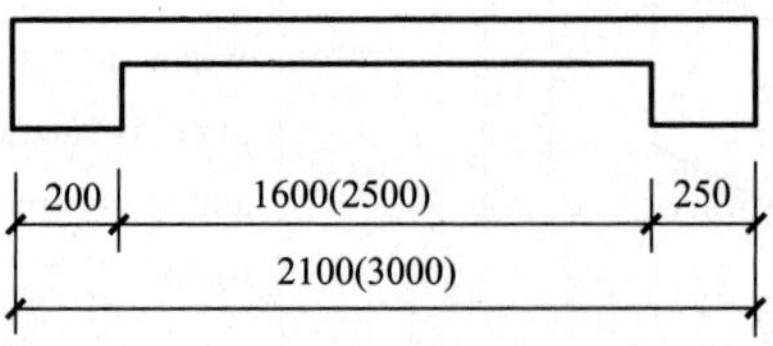	As for two similar components with different dimension figures, different dimension figures may be marked in brackets

Continued

Contents	Examples				Making methods
Some similar components	![diagram with dimensions 400, 600, a, b, c]				If some dimension figures are different, a table shall be used to show specific sizes
	Code	a	b	c	
	Z-1	200	200	200	
	Z-3	250	450	200	
	Z-1	200	450		

2.2 Common Drawing Instruments, Tools, and Their Usages

To improve the quality and plotting speed of engineering drawings, it is necessary to properly use drawing instruments and tools, such as the drawing board, t-square, set square, pencil, needle pen, compasses, beam divider, French curve, and template. This section will introduce some common drawing instruments and tools briefly.

The drawing board is used to fix and plot drawings. There are various drawing boards with different specifications. The selection of the drawing board depends on the drawing sheet and the drawing board should be larger than the drawing sheet. The surface of the board should be smooth. The working side (left side) should be selected first and the drawing board should be straight, as shown in Figure 2.2.1.

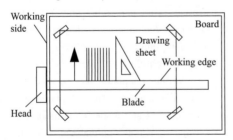
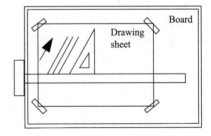

Figure 2.2.1 Drawing board and t-square

1. T-square

The t-square is used together with the drawing board to draw horizontal lines. They can also be used to draw vertical lines or oblique lines with different angles of 15°, 30°, 45°, 60°, 75°, and 105° with the aid of set squares. It consists of a horizontal ruler and a vertical ruler, which are firmly connected. The working side is the upper side of the horizontal ruler, as shown in Figure 2.2.1.

During the drawing process, the vertical ruler should cling to the working side of the drawing board to move up and down. The lines should be drawn along the upper side of the horizontal ruler from left to right. A set of horizontal lines should be drawn from up to down. The vertical lines should be drawn from down to up. A set of vertical lines should be drawn from left to right, as shown in Figure 2.2.2.

2. Set square

The set square is used together with t-square to draw vertical lines or oblique lines with different angles of 15°, 30°, 45°, 60°, 75°, and 105°. A pair of set squares have two components: 30°×60°×90° and 45°×45°×90°. They may be used to draw parallel lines and vertical lines in any direction, as shown in Figure 2.2.3.

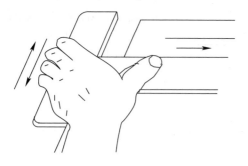
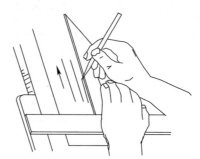

Figure 2.2.2 Drawing horizontal lines and oblique lines by t-square

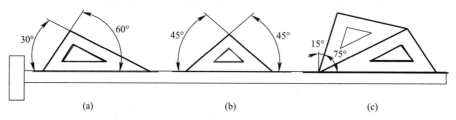

Figure 2.2.3 Drawing methods of different angles of 15°, 30°, 45°, 60°, and 75°

3. Pencils and needle pens

Pencils may be classified into soft and hard pencils and there are different codes respectively indicating soft or hard pencils at the ends of the pencil-holders. For example, "B" indicates that the pencil is soft and deep. "H" indicates that the pencil is light and hard. "HB" indicates that the pencil is neither too hard nor too soft. The bigger number before "H", the harder the pencil. The bigger number in front of "B", the softer the pencil.

Pencils should be selected according to different purposes. H and 2H pencils are generally used to draw draft lines. B and 2B pencils are usually used to deepen lines and HB pencils are used for writing.

Needle pens are widely used for inking. Its nib is a needle and can absorb carbon black ink for plotting lines. It is simple and convenient for drawing. Needle pens with different diameters can draw different lines with different widths.

Before using a needle pen, it is necessary to check whether the nib is fluent. The nob should be perpendicular to the paper during writing. The pen tube and nib should be cleaned completely when they will not be used for a long time.

4. Template

Various templates may be used to improve drawing efficiency. The templates are mostly sheet plastics and made into various shapes according to actual applications. Common templates include the multiple-use template, curved template, erasing shield, professional templates, etc. Figure 2.2.4 shows common drawing templates. Figure 2.2.5 shows the curved template.

Graphic patterns in drawing templates include some drawing symbols such as the elevation symbols, north arrow, arrows, etc.

A curved template may be used to plot non-circle curves. There are various graphic

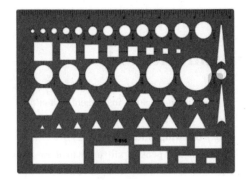

Figure 2.2.4 Templates

Figure 2.2.5 Curved template

patterns in a curved template. Its usage is introduced as follows. Firstly, enough points in curves are determined and then connected with the pencil. Then, with the aid of the curved template, several curves which coincide with certain curves in the template are connected.

5. Compasses and divider

The compass is used to draw circles or arcs. Its foot may be replaced. Needle foot, pencil foot, and ink line foot may be used. After the needle foot is installed, it becomes the divider, which is mainly used to divide a line or arc. The pencil foot may be used to plot pencil drawings and the ink line foot may be used to draw ink drawings. The feet of the divider should be always adjusted to make the drawing point perpendicular to the paper. Figure 2.2.6 shows the schematic diagrams of the compass and divider.

6. Scale

The scale is a special ruler for measuring the length according to a certain proportion. Figure 2.2.7 shows different scales. For example, if the scale of 1 : 100 is used to draw a length of 3000mm, it just needs to find 3.0m on the scale of 1 : 100. The length from 0 to 3.0m in the ruler is the corresponding length in the drawing.

Besides the above drawing instruments, drawing paper, abrasive paper erasers, adhesive tapes, pens, inks, blades, and other drawing tools should be prepared before drawing.

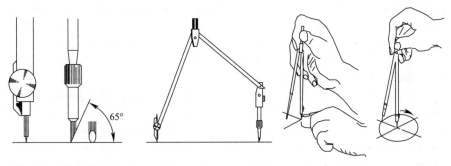

Figure 2.2.6 Usages of the compass and divider

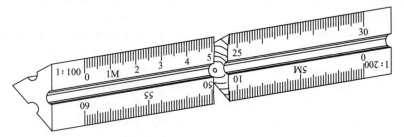

Figure 2.2.7 Scales

2.3　Drawing Steps and Methods

To improve the quality and efficiency of drawings, it is necessary to use drawing instruments and tools properly. In addition, drawing methods and steps are also important.

Drawing steps and methods may be determined according to the contents of a drawing and the habits of the drawer. The general drawing steps include drawing draft lines, proofreading, coloring, reviewing, and verifying.

Before drawing, drawing tools need to be prepared, including the drawing board, set square, t-square, pencil, eraser, blade, compass, template, and adhesive plaster. The tools and hands should be kept clean to ensure the drawing quality. A quiet environment is required during the drawing process. The drawing board should be slightly inclined. Except for t-square and set square, other items should not be placed on the board. During the drawing process, the tools and instruments should be placed in proper positions.

1. Preparation

Preparation for drawing includes sheet selection, paper cutting, and fixing, etc. The sheet size should be selected firstly based on the amount, content, size, and scale of a drawing. The sheet size should be selected based on the comprehensive consideration of the place of dimension marks and texts. The whole sheet should be well balanced.

Drawing paper should be slightly larger than the drawing sheet and should not be cut along sheet boundaries.

The size of the drawing board should match the drawing paper. If drawing paper is relatively small, it should be fixed on the left and lower part of the drawing board. The majority of the t-square should be within the drawing board when t-square is used to draw a horizontal line at the bottom.

Drawing tools and instruments should be placed in proper positions before drawing.

2. Drawing lines

H and 2H pencils are used to draw draft lines.

The drawing steps are introduced as follows:

1) Frame lines should be plotted firstly, including format lines, drawing frame lines, and title column.

2) Arrangement of drawing: Various patterns should be arranged well in the whole drawing based on the consideration of dimension marking, text notes, and intervals between drawings.

3) Drawing: The drawing order should be fully considered. For instance, the plan drawing should be drawn firstly in the architectural construction drawing and then the elevation drawings and profile drawings are drawn based on the plan drawing. In each drawing, the axis and center lines are drawn firstly and then the main contour lines are drawn. Finally, the details such as the boundary lines, dimension lines, start-stop symbols, and other symbols are drawn.

4) Checking: Coloring work may be completed after proofreading the lines.

3. Deepening the lines

Pencils or ink may be used to deepen the lines. Thin lines will be drawn first and then bold lines will be deepened from the left to right, and from top to the bottom.

4. Text notes and various symbols

According to font requirements, the grids shall be drawn, and then drawing name, scale, dimension figures, necessary text notes, title column, etc. are added. After checking, unnecessary lines shall be wiped off to keep the drawing clean.

Chapter 3
Architectural Construction Drawings

Architectural construction drawings (Hereinafter referred to as "ACD" for short) are the drawings accurately drafted by designers according to the requirements of construction and owners with the orthographic projection method in related national construction standards. All the design items of a building, such as exterior styling, interior layout, building construction, interior, and exterior decoration, and components and fittings of each part shall be drawn in detail in the drawings. A set of construction drawings of a house usually include a drawing list and general instructions of building construction, construction site layout plan, floor plan, elevation drawing, section drawing, and detail drawing. In this chapter, taking a project case as an example, revealing the contents, drawing methods, and steps of the above architectural construction drawings are introduced.

3.1 Overview

1. Types, elements, and functions of buildings

Construction is a generic term for buildings and structures. Buildings are important places for human beings to produce and live. Buildings can be classified into productive buildings and non-productive buildings in terms of various functions. Non-productive buildings refer to civil buildings, whereas productive buildings refer to industrial or agricultural buildings. Civil buildings include residential buildings and public buildings. For example, houses, dormitories, and apartments belong to residential buildings, whereas schools, hospitals, stations, hotels, airports, and gymnasiums belong to public buildings. Industrial buildings include factories and warehouses, etc. Agricultural buildings include granaries, feed yards, etc.

According to the types of bearing structures, buildings can be classified into the following six structures. Firstly, masonry structures refer to buildings whose bearing structure is the masonry structure built by blocks (such as common clay bricks) and mortar masonry. Secondly, frame structures refer to buildings with the bearing structure of frame system composed of frame beams and frame columns. Thirdly, shear wall structures refer to buildings with vertical and horizontal reinforced concrete walls as bearing structures and beams and slabs as the floor system or roof. Fourthly, in frame-shear wall structures, based on frame structures, several reinforced concrete walls are set along with the vertical and lateral directions of the building's inner frame as bearing shear walls. Therefore, frame-shear wall structures are the combination of frame structure and shear wall structure. Fifthly, tube structures refer to buildings with the bearing structure of a tube made up of reinforced concrete walls. Sixthly, other structures include shell structure, grid structure, and suspension-cable structure and are often used in long-span structures.

According to the number of stories, buildings can be classified into low-rise buildings with 1–2 stories, multi-storey buildings with 3–6 stories, and high-rise buildings exceeding a certain height or several stories. In the *Code for Fire Prevention of Buildings* (GB 50016—2014), residential buildings with 10 and over 10 floors and other civil buildings with the height of over 24m (except spacious buildings with a single floor, such as gymnasiums and movie theatres) are classified as high-rise buildings.

The buildings mentioned above have different features in terms of the construction scale, building functions, types of bearing structures, types of room combination and space division, building appearances, architectural elements, etc. Nevertheless, these buildings have the similar main elements, such as foundation, wall (column), beam, floor (ground and roof), staircase, door, and window. In addition, related elements also include awning, balcony, step (ramp), rain pipe, sill, ditches (or apron), plinth, and other decorative structures. Among them, the elements of the building structure are called components, such as foundation, wall, column, beam, and floor. Other elements with special functions are called fittings, such as doors and windows. The names and positions of typical building elements are shown in Figure 3.1.1. Some elements are introduced below.

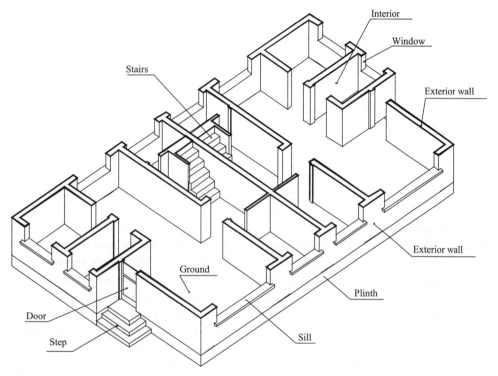

Figure 3.1.1 Basic elements of a house

Foundations are the part of the upper structure contacting the subsoil, supporting the weight of the whole building, and transferring the whole load to the subsoil.

According to the bearing condition, the walls can be classified into two types: load-bearing wall which directly bears and transfers loads (such as the masonry walls in the masonry structures and reinforced concrete wall in the shear wall structures) and non-bearing wall which is used to separate rooms (such as in-filled walls in frame structures). According to the location, walls can be classified into lateral walls, longitudinal walls, exterior walls, and interior walls.

Columns are the vertical load-bearing components that transfer the upper load to the foundation, whereas beams are the horizontal load-bearing components that transfer the upper load from the slab to columns or walls.

Floors divide the interior structures of buildings into several stories vertically and transfer their upper loads to columns or walls via supporting beams. A roof is the load-bearing structure on the top of a building, bears its upper loads, and transfers them to beams. A roof plays the role of shelter and resists wind, sand, rain, snow, and external heat.

Stairs are the component for vertical transportation and evacuation. The main function of doors is to connect rooms with transportation elements. Windows are used for lighting and ventilation. Rain pipe, ditches, and aprons are used for draining. A plinth is to protect the root of the exterior wall.

2. Steps and contents of construction engineering design

In general, the construction process of a new building can be divided into two phases, namely, the design stage and the construction stage. In the design phase, a complete set of drawing files are completed by the design party according to the requirements of owners and the global design, essential local details of the building and elements are expressed comprehensively. In the construction phase, these drawing files are used as guidelines for all the construction activities.

The design phase for civil construction engineering is generally divided into three sub-stages, namely, schematic design, preliminary design, and construction drawing design. As for the construction engineering projects with simple technological requirements, after the schematic design is approved, the construction drawing design can be directly started and the preliminary design can be omitted under the permission of the competent department.

1) Schematic design

The schematic design is completed in the initial stage of the design process and mainly involves the primary ideas about the overall layout and spatial organization of the building. The drawings completed in this stage are simple and known as conceptual drawings. In a successful schematic design, many factors should be comprehensively considered, including the requirements of owners, the conditions of construction technologies, investment, and architectural aesthetics.

The files of schematic design mainly include design specifications (design specifications of various disciplines and investment estimation), general layout and architectural design drawings (plan, elevation, section, etc.), and files required in the design commission or design contract and such as perspective drawings, aerial views, models, and related economic indicators. The drawings completed in the schematic design stage are the documents provided to the competent department for approval.

2) Preliminary Design

Based on the approved schematic design drawings, various technical issues should be further addressed, such as the mold and layout of components. In addition, the technical contradictions among different professions should be solved to carry out technical and economic comparisons and necessary calculations. The preliminary design stage is the final stage of various technical issues and is also called the technical design stage.

The drawings completed in the preliminary design stage are almost the same as those completed in the schematic design stage except that more details are included in the former. The approved preliminary design drawings and documents are the foundation for preparing the construction drawings, purchasing main materials and equipment, and allocating project funds.

3) Construction drawing design

Based on the preliminary design documents approved by the competent department, a full set of drawings reflecting the overall and partial building styles and related technical materials should be prepared in the construction drawing design stage. To meet the specific requirements of construction activities, more details should be provided in these drawings as possible.

The documents completed in the construction drawing design stage include design drawings, specifications, and calculation sheets of architectures, structures, and equipment. The construction design drawings are the basis for subsequent construction activities. Construction drawings and preliminary design drawings are almost the same in terms of the diagram principle and drawing methods, but the number or details of drawings are different.

According to the disciplines, the contents of building design include architectural design, structural design, and equipment design.

(1) Architectural design

The architectural design should be conducted based on the overall planning and design specifications of a building. Engineers determine the designs of the plan and spatial relations, as well as the exterior style of the building. A complete set of construction drawings of the building are drafted based on the comprehensive consideration of the following factors: the construction environment, functions, the bearing structures, building economy, and architectural aesthetics.

(2) Structural design

According to the architectural design completed in the previous stage, an appropriate load-bearing structure plan is determined by structural engineers. In the structural design stage, the layout of bearing structures and the design of bearing components are determined and a full set of structural construction drawings are completed.

(3) Equipment design

In the equipment design stage, water supply and drainage, electrical lighting, heating, and ventilation are designed and a full set of the equipment construction drawings are drafted.

Designers of architectural, structural, and equipment backgrounds are required to cooperate in the design of a building. These designers are required to complete the design items in their own major and cooperate to constitute the whole design content of the building. The drawings, specifications, and calculation sheets of the three disciplines constitute a full set of construction engineering design materials for construction activities.

3. Classification and order of construction drawings of buildings (CDB)

According to various disciplines and functions, CDB can be classified into architectural construction drawings, structural construction drawings, and equipment construction drawings. A full set of building construction drawings generally includes a drawing list, general construction instructions, architectural construction drawings, structural drawings, and equipment construction drawings.

1) Architectural construction drawings

Architectural construction drawings are called "ACD" for short and prepared after the schematic design stage or preliminary design stage. ACD includes some required items, such as the overall layout, interior layout, exterior decoration, detailed structures, some fixed facilities, and construction requirements.

Architectural construction drawings are used as the basis for the construction of basic architectural elements, such as foundations, walls, windows, doors, and interior and exterior decorative structures, which are considered in preparing budgets and construction organization plans.

A full set of architectural construction drawings generally include a drawing list, general construction instructions, basic drawings indicating overall contents (such as site plan drawings, architectural plan, architectural elevation, and architectural section), detail drawings indicating the detailed sizes and materials of various components or parts (such as stairs, doors, windows, toilets, bathrooms, and decorative components), and the tables of doors and windows.

2) Structural construction drawings

Structural construction drawings are called "SCD" for short and mainly record the layout, types, sizes, materials, and structural practices of load-bearing components.

Structural construction drawings are the basis for the construction of structural bearing components, such as foundations, walls (or columns), beams, and slabs.

A full set of structural construction drawings generally include drawing contents and general instructions of structural drawings, basic drawings (such as the foundation layout and structural plan drawings), and detailed drawings of structural components (such as the number and layout of reinforcement bars for shear walls, columns, beams, slabs, and stairs).

3) Equipment construction drawings

Equipment construction drawings are called "ECD" for short and mainly record the layout, trend, and structural installation requirements of drainage, heating, ventilating, electric appliance, etc. The equipment construction drawings generally record the layout plan, system diagram, and detailed drawing of various equipment.

According to various disciplines, architectural construction drawings can be classified into ACD, SCD, and ECD. All three kinds of construction drawings include basic drawings and detailed drawings. A simple set of architectural construction drawings is composed of several or dozens of drawings, whereas a set of complicated architectural construction drawings even include hundreds of drawings. Therefore, to look up the drawings efficiently, these drawings shall be arranged in proper order. According to *Unified Standards for House Building Drawings* (GB/T 50001—2017), architectural construction drawings should be arranged in disciplines, such as drawing contents, general drawings, building drawings, water supply, and drainage drawings, heating and air conditioning drawings, and electrical drawings. The construction drawings of different disciplines should be arranged following the primary and secondary relations and logical relations. General drawings are arranged before detail drawings; overall drawings are arranged before partial drawings; layout drawings are arranged before component drawings.

4. Building module coordination

The *Coordination Standards for Building Module* (GB/T 50002—2013) has been launched by the Chinese government in order to promote the large-scale industrial production of building products, components, fittings, and assembling unites and improve the versatility and interchangeability of components made from different materials, forms and methods. Based on this standard, the design speed can be accelerated, thus increasing the efficiency and decreasing the cost.

Modules are defined as the units with the selected size and can be classified into basic modules and derivative modules.

According to the *Coordination Standards for Building Module* (GB/T 50002—2013), the value of the basic module is 100mm with the symbol of M. In other words, 1M equals 100mm. The modulated sizes of the whole building, a part of the building and building assembles should be the multiples of the basic module.

1M is also used as the horizontal and vertical basic modules. A horizontal module mainly refers to the horizontal dimension such as the width of doors and windows and ranged from 1M to 20M. The vertical module always refers to the height of a storey or windows and doors and ranges from 1M to 36M.

Derivative modules are derived from basic modules and can be classified into enlarged modules and reduced modules. The cardinal numbers of enlarged modules are 2M, 3M, 6M, 12M, 15M, 30M, and 60M and the corresponding sizes are respectively 200mm, 300mm, 600mm, 1200mm, 1500mm, 3000mm, and 6000mm. The cardinal numbers of reduced modules are M/10, M/5, and M/2 and corresponding sizes are respectively 10mm, 20mm, and 50mm.

Horizontal enlarged module series shall adopt $2n$M and $3n$M (n is a natural number), which are mainly used for the column distance, the depth or width of rooms, the dimensions of components and fittings and the widths of doors and windows.

Vertical enlarged module series shall adopt nM (n is a natural number), which is mainly used in the building height, storey height, and the height of doors, windows and holes.

The series of reduced modules are introduced below. 1/10M series shall be upgraded from 1/10M to 2M with the step of 10mm; 1/5M series shall be upgraded from 1/5M to 4M with the step of 20mm; 1/2M series shall be upgraded from 1/2M to 10M with the step of 50mm. Sub module series of 1/10M, 1/5M, and 1/2M shall be applied in gaps, structural joints and the sections of components and fittings.

In the dimension coordination of different building types and parts, the size ranges shall be reduced as possible to allow the flexible superposition and segmentation.

5. Standard atlas

In order to accelerate the design process and improve the drawing quality, various commonly used buildings, components and fittings are designed and plotted in a set of standard construction drawings according to different specification standards and module coordination requirements in international standards. These drawings are called standard atlas or collective drawings and can be bound together in a volume to obtain a standard drawing album or common drawing album.

There are two types of standard atlas: the building component drawing atlas indicated by G and the building fitting drawing atlas indicated by J. For example, the common drawings in Jiangsu Province contain **Collective Drawings of Plastic Doors and Windows** with the classification number of "Su J30—2008" and **Collective Drawings of Common Joints of Building Structure** with the classification number of "Su G01—2003".

The standard atlas shall be selected according to its applicable scope. Standard atlas or general atlas approved by national authorities can be used nationwide, whereas those approved by the authorities of provinces, cities or autonomous regions are only applicable to corresponding areas.

3.2 Regulations of Architectural Construction Drawings

Architectural construction drawings should be drafted and read according to the orthographic projection principle, **Unified Standards for House Building Drawings** (GB/T 50001—2017). The overall plan drawings should be drafted and read according to **Overall Drawing Standards** (GB/T 50103—2010). Elevation drawings, section drawings and detail drawings should be drafted and read according to **Standard for architectural drawings** (GB/T 50104—2010). In this section, the provisions for the scales and lines of architectural drawings and the legends and dimension notations of components and fittings in **Standard for architectural drawings** (GB/T 50104—2010) are introduced. Moreover, the locating axes and legends of construction materials in **Unified Standards of House Building Drawings** (GB/T 50001—2017) are also introduced.

1. Scales

According to **Standard for architectural drawings** (GB/T 50104—2010), the scales in building drawings shall meet the scale requirements in Table 3.2.1.

Scales of building drawings

Table 3.2.1

Names of Drawings	Scales
Plan, elevation and section drawings of buildings	1 : 50, 100, 150, 200, and 300
Partial enlargement drawings of buildings or structures	1 : 10, 20, 25, 30, and 50
Detail drawings of fittings and structures	1 : 1, 2, 5, 10, 15, 20, 25, 30, and 50

2. Drawing lines

Lines in architecture drawings shall comply with the provisions in Table 3.2.2. Before the drawing, the width b of bold lines should be chosen according to the conditions of drawing patterns. In general, the width of bold lines is related to the scale and the complexity of a drawing.

3. Locating axis and numbering

The locating axes in architectural construction drawings are the important basis for construction and locating. The main load bearing components such as load bearing walls and columns should be chosen as the reference position to draw the locating axis.

The provisions for locating axis in the ***Unified Standards for House Building Drawings*** (GB/T 50001—2017) are introduced as follows.

Locating axis shall be drawn with thin dotted line and numbered and the series number shall be written within the circle at the end of the axis. The circle shall be drawn by thin solid line according to the diameter of 8-10mm. The center of the circle shall be positioned on the extension line of the locating axis.

Selection of drawing lines Table 3.2.2

Names	Line Types	Line width	Applications
Bold solid line	————	b	Contour lines of main building structures (including components and fittings) sectioned in the plan and section drawings; outer contour lines of building elevation drawings or indoor elevation drawings; contour lines of the main bodies sectioned in the detail drawings; outer contour lines in the detail drawings of building elements and fittings; section symbols of plan, elevation and section drawings
Semi-bold solid line	————	$0.7b$	Contour lines of secondary structures (including components and fittings) sectioned in plan and section drawings; contour lines of building elements and fittings in plan, elevation and section drawings; general contour lines in the detail drawings of structures, elements and fittings
Middle solid line	————	$0.5b$	Drawing lines, dimension lines, dimension boundaries, indexing symbols, elevation symbols, and guide lines, paint lines, insulating layer lines, elevation boundaries of grounds and walls in detail drawings of the material procedure
Thin solid line	————	$0.25b$	Filled lines of legends, furniture lines, pattern lines, etc

Continued

Names	Line Types	Line width	Applications
Semi-bold dash line	----------------	0.7b	Invisible contour lines of detail drawings of structures, components and fittings; contour lines of lifters (cranes) in the plan drawings; contour lines of buildings to be expanded
Middle dash line	----------------	0.5b	Projecting line's invisible contour lines with line width less than 0.7
Thin dash line	----------------	0.25b	Invisible contour lines of filled lines of legends and furniture lines
Bold single dot long line	—·—·—·—·	b	Track lines of lifters (cranes)
Thin single dotted long line	—·—·—·—	0.25b	Centerlines, symmetrical lines, and locating axis
broken line	——⌐⌙——	0.25b	Disconnected boundary indicating partial omission
Wave line	∿∿∿	0.25b	Disconnected boundary indicating partial omission; disconnected boundary of curve components; disconnected boundary of tectonic levels

Note: The width of horizontal lines is 1.4b (extra bold line).

The serial number of locating axis in a plan drawing should be indicated at the bottom and the left of the drawing. The horizontal serial number should use Arabic numerals and be written from the left to the right. The vertical serial number should use Latin capital letters and be written from the bottom to the top (Figure 3.2.1). To avoid confusion, capital Latin letters

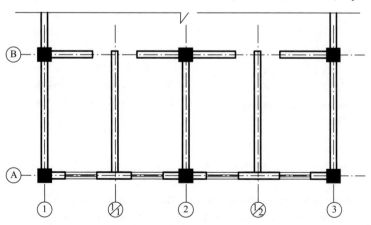

Figure 3.2.1 Numbering order of the locating axis

(I, O, or Z) can not be used individually as the serial number of axis. If single Latin letters are not enough for numbering, double letters or letters with the number subscript can be used, such as A_A, B_A …… Y_A, A_1, B_1 …… Y_1.

In complicated plans, a partition number (Figure 3.2.2) may be used to denote a locating axis and the serial number should be denoted as "partition number-serial number". Both Arabic numerals and Latin capital letters may be used as the partition number.

As for the building with a circle or arc plan, the radical axis should be numbered with Arabic numerals from the left bottom of a drawing in counter-clockwise orders. The circular axis should be numbered with capital Latin letters from outside to the center of the circle, as shown in Figure 3.2.3.

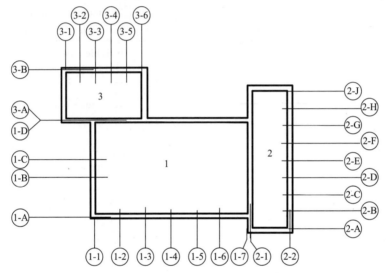

Figure 3.2.2 Partition number of a locating axis

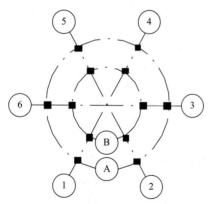

Figure 3.2.3 Numbering order of locating axis in a circle plan drawing

The locating axis in breaking-line plan drawings should be numbered, as shown in Figure 3.2.4.

When a detail drawing is applicable to several axes, the serial number of each axis should be indicated (Figure 3.2.5). As for the locating axis in general detail drawings, only a blank circle without numbering is required.

Secondary partial load-bearing components can be denoted with additional locating axis (sub axis) or by marking the distance away from its nearby axis. The sequence of additional locating axis should be expressed in the form of fraction and written in accordance with the following requirements:

1) As for additional axes among main axes, the denominator indicates the serial number of former axis, whereas the numerator indicates the serial number of additional axis. $\left(\frac{2}{3}\right)$ The second additional axis after No.3 axis. $\left(\frac{3}{B}\right)$ The third additional axis after No.B axis.

2) The additional axis before No.1 or No.A axis should be denoted by adding 01 or 0A in

the location of denominator. The third additional axis before No. 1 axis. The second additional axis before No. A axis.

4. Legends of structures and elements of a building

Since buildings and structures are usually scaled down in a drawing, some building details, component forms or construction materials can not be clearly drawn realistically or literally expressed. When different structures are expressed with unified legends, they can be understood easily and clearly. Therefore, various legends are generally introduced in construction engineering drawings.

Table 3.2.3 shows the provisions and descriptions of commonly used legends of structures and components in *Standard for architectural drawings* (GB/T 50104—2010). Table 3.2.4 shows the legends and descriptions of commonly used horizontal and vertical transporting devices.

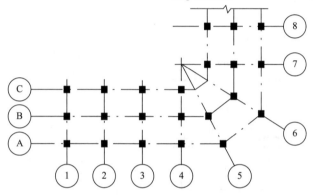

Figure 3.2.4 Numbering order of locating axis in a breaking-line plan drawing

Figure 3.2.5 Numbering order of axes in detail drawings

Legends of structures and components of a building　　　Table 3.2.3

Names	Legends	Descriptions
Stairs	Upward / Downward / Upward / Downward	The upper figure is the bottom stair plan; the middle figure is the middle stair plan; the lower figure is the top stair plan. The stairs, handrails and step numbers of the stairs should be drawn according to the actual situation

Continued

Names	Legends	Descriptions
Walls	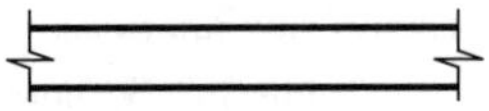	Words or legends should be added to indicate the wall materials and the table of material legends should be listed in the project design drawings
Slopes	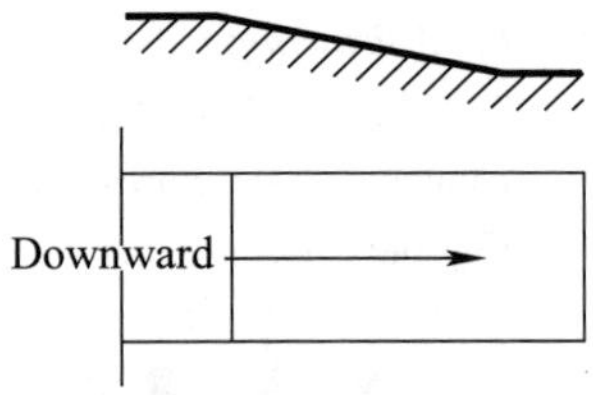 Downward (upper long slope figure); Downward (doorway slope figures)	The upper figure shows a long slope and the lower figure shows a doorway slope
Partition	 	Including partitions of lath and plaster, woods, plasterboard and metal materials; applicable to the partitions reaching the top or not
Handrail		
Holes	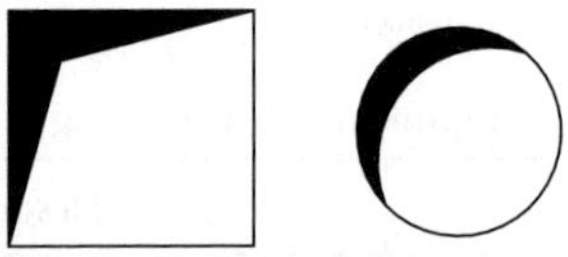	Shaded parts can be replaced by painting
Reserved holes on the wall	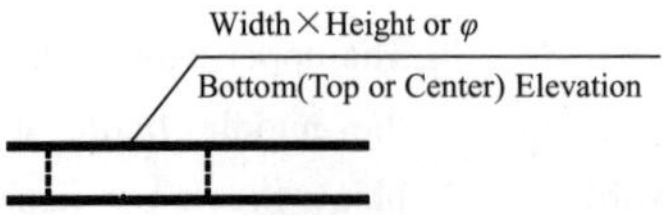 Width × Height or φ / Bottom(Top or Center) Elevation	Locating at the hole center or around. It is better to paint to distinguish the locations of walls and retained holes
Reserved grooves on the wall	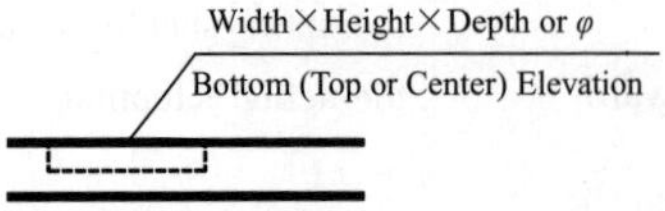 Width × Height × Depth or φ / Bottom (Top or Center) Elevation	

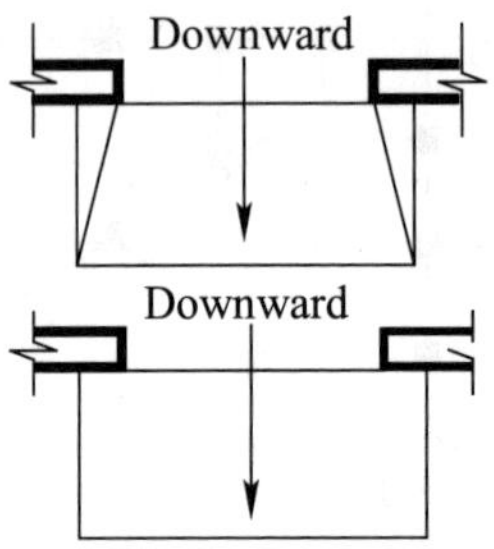

Continued

Names	Legends	Descriptions
Flues	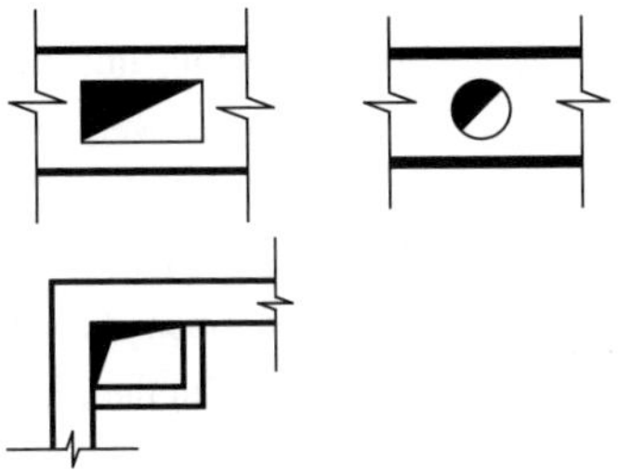	The shadow parts can be replaced by painting. Flues and walls are made from the same materials and the wall lines of their connections should be broken
Air drain	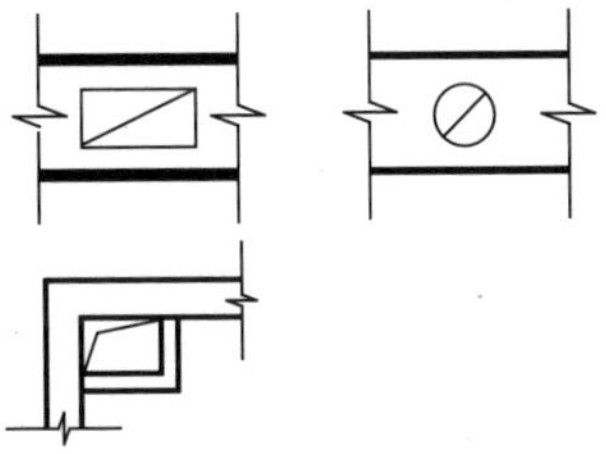	
Walls and windows	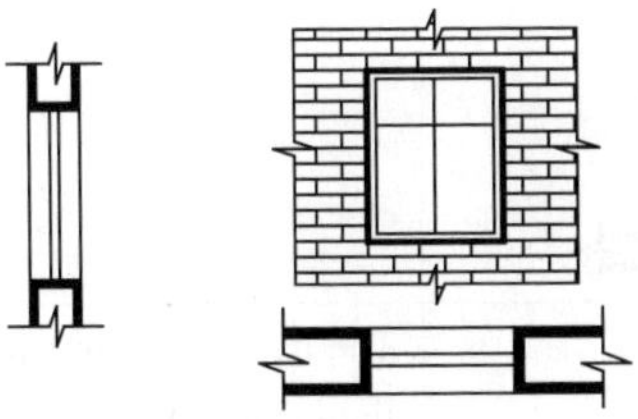	The legends in the figure are small building blocks, which should be plotted according to used materials. Those that are difficult to be drawn with legends can be indicated by words or codes on walls. Under a small scale, the windows in plan drawings and section drawing can be drafted with single bold lines
Empty doorway	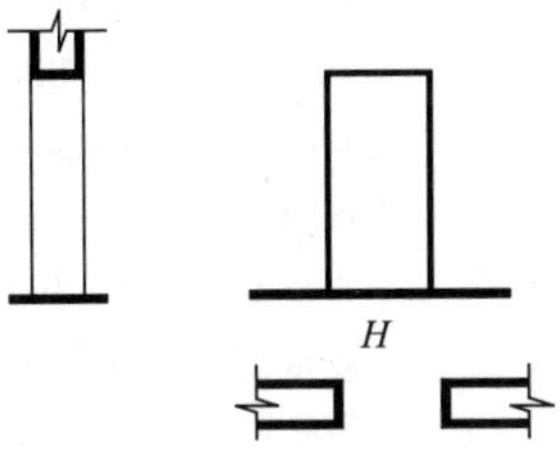	"H" refers to the height of the doorway
Single door (including flat-opening or single-sided spring)	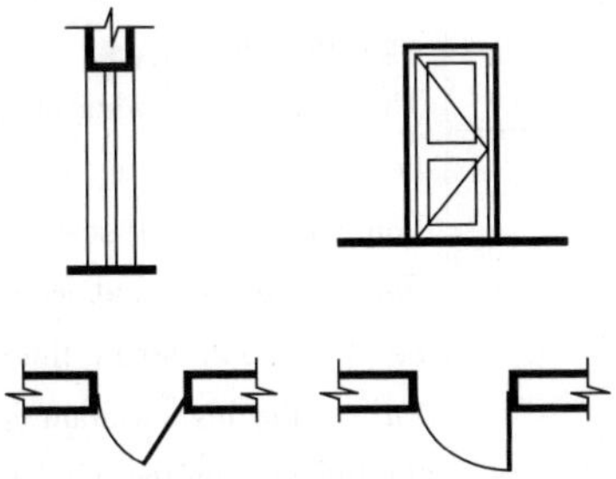	The name code of doors is M. The hinge is installed on one side of intersecting angle in opening direction line. The solid line refers to opening outwards and the dotted line refers to opening inwards. Door lines in the plan figure should be opened to form an angle of 90° or 45° and the opening arc should be drawn.

Names	Legends	Descriptions
Double door (including flat-opening or single springs)	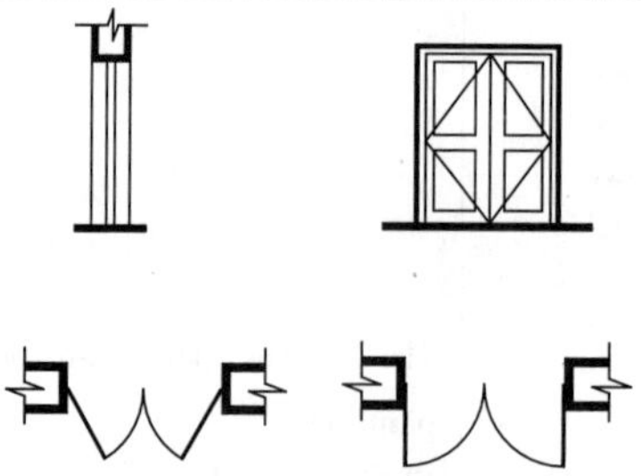	The opening line is unnecessary in the elevation drawing, but it should be indicated in the detail drawings and indoor design drawings. The elevation drawing should be drawn according to the actual situation
Sliding door		The name code of doors is M. The elevation drawing should be plotted according to the actual situation
Vertical roller shutter door	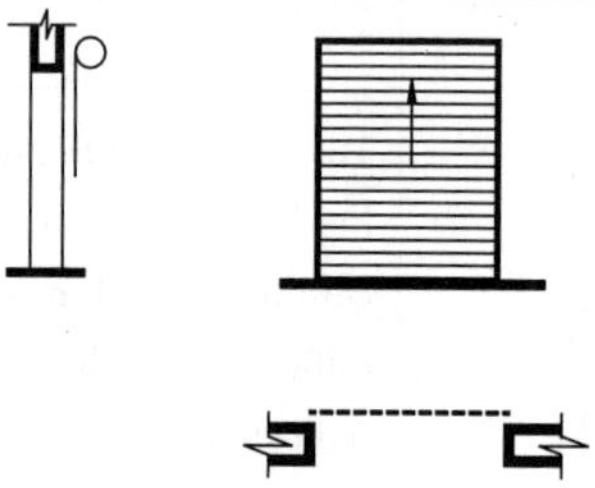	
Single-layer fixed window	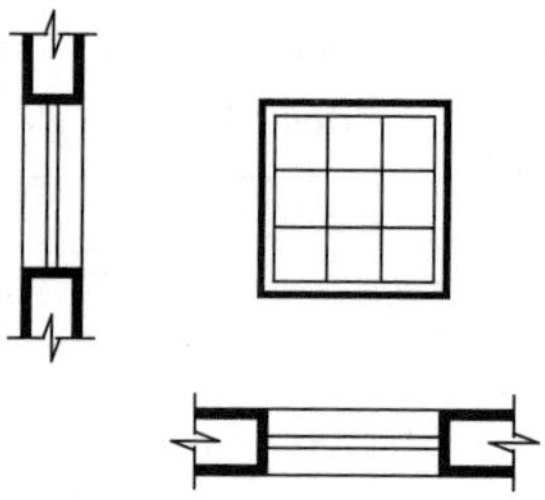	The name code of windows is C. The oblique line in the elevation drawing indicates the opening direction of windows; the solid line refers to opening outwards and the dotted line refers to opening inwards. The hinge is installed on one side of the intersecting angle in the opening direction line, but it is unnecessary in general design drawings. The dotted lines in the plan drawings and section drawings only indicate the opening way, but it is unnecessary in the design drawings. The elevation drawing of windows should be plotted according to the actual situation. Under a small scale, the windows in the plan drawing and section drawing can be plotted with single thick solid lines. "h" in the high window legend refers to the height from the window bottom to the ground of this storey
Sliding sash	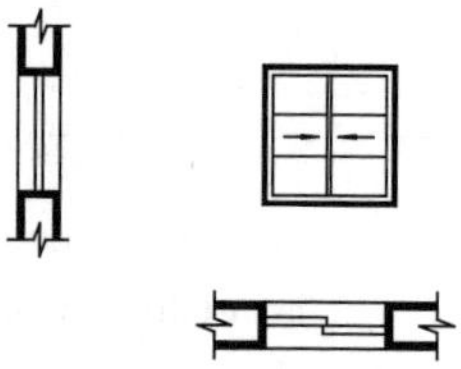	
High window	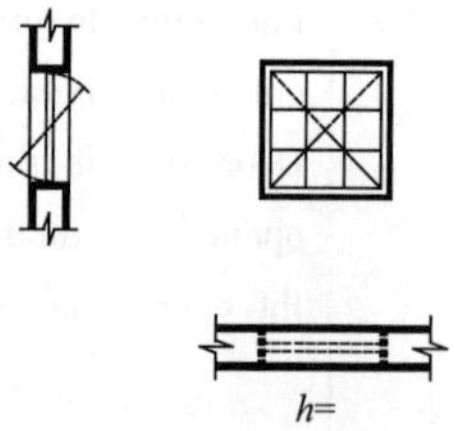 $h=$	

Legends of horizontal and vertical transporting devices   Table 3.2.4

Names	Legends	Descriptions
Elevator		The elevator type shall be indicated and the actual positions of the door and the balance hammer shall be drawn. Special types of elevators such as sightseeing elevators should be drawn according to the actual situation and this legend
Escalator	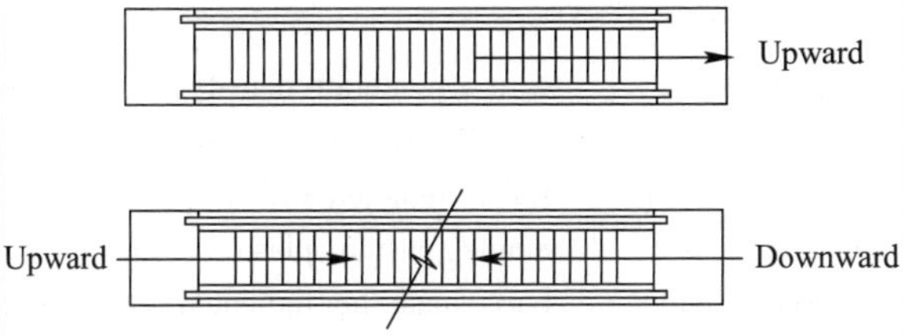 Upward / Upward — Downward	Escalator, moving sidewalk and automatic pedestrian ramp can run forward and backward. The direction of arrows indicates the designed running direction. At the bottom of arrows in automatic pedestrian ramp, upward or downward label should be added
Moving sidewalk and automatic pedestrian ramp	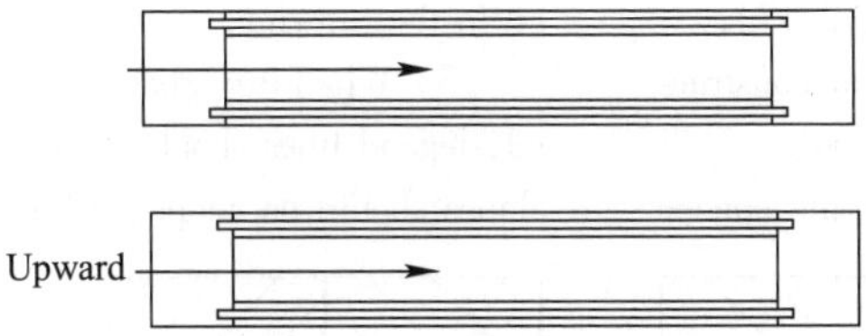 Upward	
Railway	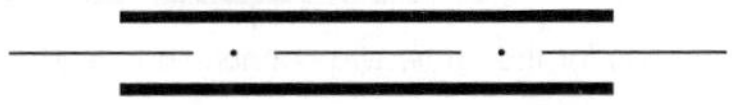	This legend is applicable to standard railway and narrow-gauge railway and rail gauge should be indicated in this legend
Bridge crane	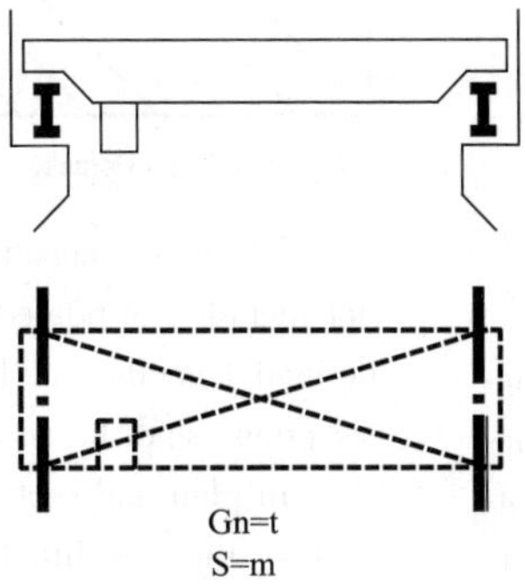 Gn=t S=m	The upper figure is the elevation (or section) drawing and the lower figure refers to the plan drawing. The legend of a crane should be drawn according to the scale. Operating house should be drawn when necessary according to the actual situation. The name of the crane, the scope of running axis and working class should be written.

Continued

Names	Legends	Descriptions
Bridge crane	$G_n = t$ $S = m$	G_n: Refers to the elevating capacity of the crane in the unit of "t". S: Refers to the span or brachium of a crane in the unit of "m"

5. Legends of construction materials

In construction engineering drawings, legends are also adopted to represent construction materials. Table 3.2.5 lists commonly used construction material legends and other legends can be found in *Unified Standards for House Building Drawings* (GB/T 50001—2017).

The provisions of the legends of construction materials are introduced as follows:

1) Legend lines should be evenly spaced with a moderate density.

2) When different types of the same material is denoted with a common legend (for example, when plasterboards in some specific parts must be indicated as waterproof plasterboards), necessary descriptions should be added in the drawing.

3) When two adjacent legends are connected, legend lines should be staggered or opposite slopes should be adopted (Figure 3.2.6).

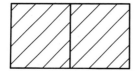 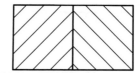

Figure 3.2.6　Drawing methods for the same two connected legends

4) A gap of not less than 0.7mm should be left between two adjacent black legends (Figure 3.2.7).

Figure 3.2.7　Drawing methods of two adjacent black legends

along the section contour lines (Figure 3.2.8).

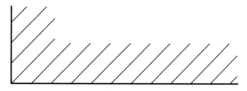

Figure 3.2.8　Partially displayed legend

5) When only one legend is used in a figure or the drawing is too small to draw construction material legends, the legends can be omitted, but literal descriptions should be provided.

6) When the construction material legend is too large, a part of the legend may be plotted

7) When construction materials that are not included in related standards are used, self-defined legends can be used, but related descriptions shall be added at a proper location.

In plan and section drawings with different scales, the simplified drawing methods of the plaster layer, ground floor and material legends should meet the following requirements:

Legends of commonly used construction materials   Table 3.2.5

Names	Legends	Descriptions
Stock brick	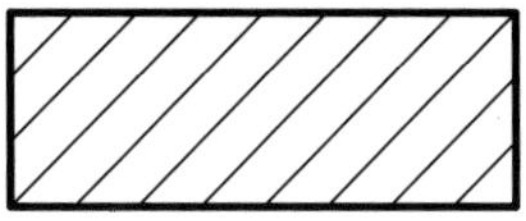	Masonries include solid brick, porous brick, building blocks, etc. When the section is too narrow to draw masonries in legend lines, they can be painted red
Hollow brick	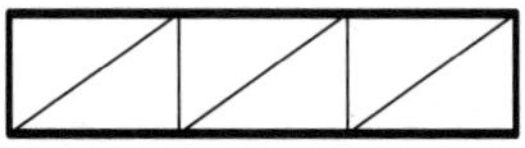	Refers to non-load-bearing masonries
Reinforced concrete	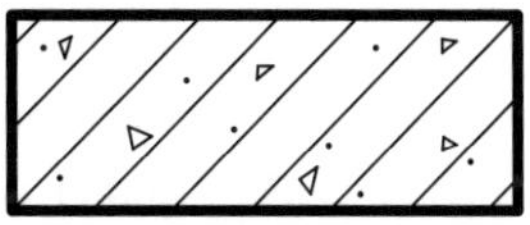	This legend refers to load-bearing concrete and reinforced concrete, including concrete of different strengths, aggregates and additives. When steel is drawn in the section drawing, legend lines are unnecessary. When the section drawing is too small to draw legend lines, it can be painted black
Concrete	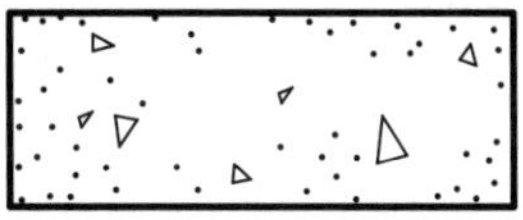	
Compacted soil	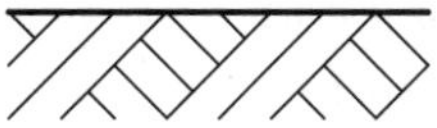	
Natural soil		Including various natural soils
Stone	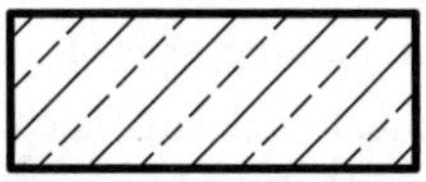	
Foamed plastic material	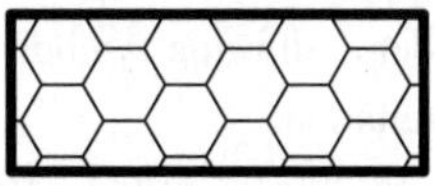	Including porous polymer materials, such as polystyrene, polyethylene, and polyurethane
Metal	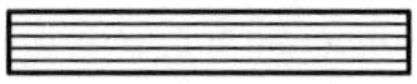	(1) Including various metals. (2) When the drawing is small, it can be painted black
Glass	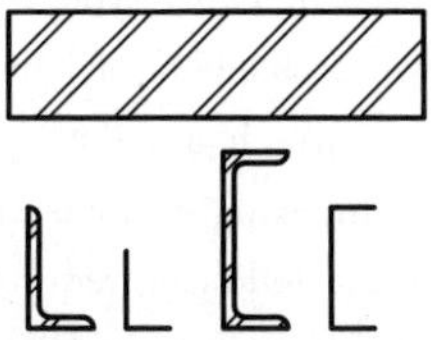	Including flat glass, frosted glass, wired glass, toughened glass, insulating glass, laminated glass, coated glass, etc

1) In plan drawings and section drawings with a scale larger than 1 : 50, the surface course lines of plaster layer, ground floor, and roof as well as material legends should be plotted.

2) In plan drawings and section drawings with a scale of 1 : 50, the surface course lines of ground floor and roof should be drawn and the surface course lines of plaster layer should also be plotted.

3) In plan drawings and section drawings with a scale less than 1 : 50, the plaster layer should be drawn, whereas the surface course lines of ground floor and roof should also be drawn.

4) In plan drawings and section drawings with a scale from 1 : 200 to 1 : 100, material

legends may be simplified (such as painting the masonry wall red and painting the reinforced concrete black), whereas the surface course line of ground floor and roof should be drawn.

5) In plan drawings and section drawings with a scale less than 1 : 200, material legends and the surface course lines of ground floor and roof may be omitted.

6. Dimensions

The dimension markings in architectural construction drawings should be clear and complete. In addition to some necessary exterior dimensions and interior detail dimensions, another type of dimension marking, elevation labeling, is also commonly used.

The requirements for elevation labeling are introduced below. In the plan, elevation and section drawings of buildings, the elevations of indoor and outdoor terraces, ground floors, underground floors, balconies, platforms, cornices, ridges, parapet walls, canopies, windows, doors, steps, etc. should be indicated. As for the flat roofs whose building elevations are difficult to be indicated, the structural elevations can be indicated along with descriptions. As for the flat roof with a structural slope, the roof elevation may be indicated at the bottom of the structural layout together with slope descriptions. As for the flat roof with a roof truss, the elevation of its bottom chord lay-down point or capital should be indicated. As for the plant building with cranes, the elevations of the rail top, the lower edge of roof truss, beam bottom and board bottom should be indicated. The gauge size of the rail suspension crane should also be indicated in the unit of "m".

The heights and elevations of ground floors, underground floors, balconies, platforms, cornices, ridges, parapet walls, steps, etc. should be indicated according to the following requirements:

1) In plan drawings and detail drawings, the elevations of finished surfaces shall be indicated.

2) In elevation drawings, section drawings and detail drawings, the elevations of finished surfaces and the size in the height direction shall be indicated.

3) As for other parts, the sizes and elevations of rough surfaces shall be indicated.

4) When the locating sizes of different parts in plan drawings are indicated, the locations of parts may be indicated together with the distance to the most adjacent axis. To mark the locating size of the parts in elevation drawings, the dimensions may be denoted together with the elevation of corresponding storey.

7. Indexing symbol and detail drawing symbol

When the structure of a certain part or the configuration between components in a global drawing can not be clearly demonstrated and additional detail drawings are required, an indexing symbol shall be added in order to correlate the global drawing with the detailed drawing. In other words, an indexing symbol is added in the global drawing to index the detail drawing and a detail drawing symbol is added in the detail drawing to index the global drawing. The indexing symbol shall be consistent with corresponding detail drawing symbol for the convenience of consultation.

In the ***Unified Standards for House Building Drawing*** (GB/T 50001—2017), the requirements for indexing symbols are introduced below.

The indexing symbol is composed of a circle with a diameter of 10mm and a horizontal diameter, which are both plotted with thin solid lines. Indexing symbols shall be drawn according to the following requirements.

1) If an indexing detail drawing and an indexed detail drawing are arranged in the same drawing sheet, an Arabic numeral should be added in the upper semi-circle of the indexing symbol to indicate the serial number of this detail drawing and a horizontal thin solid line should be drawn in the lower semi-circle: $\dfrac{3}{-}$. The symbol indicates that the detail drawing with the

serial number of 3 and the indexed drawing are in the same drawing.

2) If an indexing detail drawing and an indexed detail drawing are not arranged in the same drawing sheet, an Arabic numeral should be added in the upper semi-circle of the indexing symbol to indicate the serial number of this detail drawing, and an Arabic numeral should be added in the lower semi-circle to indicate the serial number of the drawing sheet containing the detail drawing: $\frac{3}{JS10}$. The symbol indicates that the detail drawing with the serial number of 3 is in the drawing sheet with the serial number of JS10.

3) If the indexed detail drawing is plotted in the form of the standard drawing, the serial number of this standard drawing should be written above the extension line of the horizontal diameter of this indexing symbol: $\frac{3}{JS10}$ J103. The symbol indicates that the detail drawing is included in the standard drawing album with the serial number of J103.

4) If the indexing symbol is used to index the sectioned detail drawing, the sectioned line should be drawn on the section part with extension line to mark the indexing symbol. The extension direction is the projection direction, as shown in Figure 3.2.9.

The symbol of a detail drawing indicates the location and serial number. The circle of the symbol in a detail drawing shall be drawn in thick solid lines with the diameter of 14mm. The detail drawing should be numbered according to the following requirements.

1) When the detail drawing and the indexed drawing are in the same drawing sheet, an Arabic numeral should be indicated in the detail drawing indexing symbol to show its serial number: (3). The symbol indicates that the detail drawing with the serial number of 3 and the indexed drawing are in the same drawing sheet.

2) When the detail drawing and the indexed drawing are not in the same drawing sheet, a horizontal diameter shall be drawn with a thin solid line in the detail drawing symbol. The serial number of detail drawing is added in the upper semi-circle, whereas the serial number of the indexed drawing sheet is added in the lower one: $\frac{3}{JS01}$. The symbol indicates that the detail drawing with the serial number of 3 and the indexed drawing with the serial number of JS01 are not in the same drawing sheet.

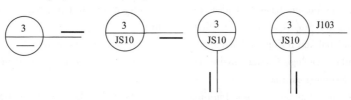

Figure 3.2.9 Indexing symbols of sectioned detail drawings

3.3 Drawing List and General Construction Instructions

The **Deep Regulations** mentioned later in this section and this book refers to the **Deep Regulations on the Compilation of Architectural Engineering Design Documents** (Version 2016).

The drawing list is an index file of construction drawings. The drawing list of construction drawings shall be prepared for one discipline of each sub-item. The same drawing list shall contain only one sub-item. According to the **Deep Regulations**, the drawing list shall be arranged in the front of construction drawings and not be included in the serial number of drawings. Some organizations are used to arranging the drawing list and general construction instructions in the same drawing. The column of the drawing list usually includes the serial number, drawing type, drawing number, drawing title, drawing sheet and remarks. The drawing number and drawing type in the drawing list shall be consistent with those in the corresponding drawing.

As necessary supplementary files, the general construction instructions are mainly the general literal descriptions of the project and shall cover the overview, design basis, materials and practices, construction and production precautions as follows.

Firstly, project overview usually includes the project name, project site, construction unit, building area, storey height, building grade, number of floors, civil air defense engineering grade, seismic fortification intensity, and the main types of structures. The project overview of the small and technically simple projects can be simplified.

Secondly, the design basis includes approval documents for project design or preliminary design, national and regional related specifications and atlas, geological data, hydrological data and meteorological data.

Thirdly, design standards include building standards, structural load grades, seismic requirements, heating and ventilation requirements, lighting standards, etc.

Fourthly, the relationships between the relative elevation and the absolute elevation are also included in the general construction instructions.

Fifthly, engineering practices should cover all parts of the construction materials and construction practices within the scope of this design, such as wall moisture barrier, roof, floor, ground, interior and exterior walls, plinth, water apron, stairs, and slope. Three expression forms are usually adopted: text description, referencing and numbering standard atlas, and detail drawings with indexing symbols.

Sixthly, the descriptions of special construction requirements are also included in the general construction instructions, such as fire prevention, corrosion prevention, and radiation prevention.

Seventhly, the descriptions of the adopted new materials and new technologies are also included in the general construction instructions.

An example of the general construction instructions of a laboratory building in a middle school is provided below.

CASE STUDY: General construction instructions of a laboratory building in a middle school

1. Project overview

1) Project Name: laboratory building of *** Middle School.

2) Construction Unit: *** Middle School.

3) Total building area: 2,500m^2.

4) The fire resistance rating of this project is Grade II, and the seismic fortification intensity is Degree VII.

5) The service life of this project is 50 years.

6) Structure type: masonry concrete structure.

2. Design guidelines

1) The fixed-point planning map provided by ***.

2) The preliminary design document of *** Middle School.

3) The engineering design of *** Middle School laboratory building approved by *** City.

4) The current national regulations on design and relevant regulations on architectural design promulgated by *** City in *** Province.

3. Absolute elevation

The elevation of ±0.000 is equivalent to the

40.55m absolute elevation of the Yellow Sea.

4. Engineering finishes

1) Moisture prevention for wall foundation and underground excavation wall

The cement mortar (20mm thick, 1 : 2) mixed with 5% waterproofing agent should be completed at the elevation of −0.060m.

The prevention of termites shall be performed after Party A contacts with the termite control center.

2) Walls

Exterior wall: 370mm MU10 brick wall.

Interior wall: 370/240mm MU10 brick wall.

The doorknob is 240mm unless stated otherwise.

3) Ground

Ground 1: the construction practices of moisture-proof terrazzo ground can be obtained from Su J9501——ground 11 (waterproof material of polyurethane, copper separator, and the spacing of 1000 for indoor use and corridor).

4) Floor

Floor 1: the construction practices of terrazzo floor can be found in Su J9501——Floor 5 (copper separator and the spacing of 1000).

5) Dado

Dado 1: the construction practices of light green brick with dado 900 can be found in Su J9501——Dado 9.

6) Interior walls

Interior walls: the construction practices of composite mortar plastering wall can be found in L96J002——Interior Wall B.

Cement plaster angle staff of indoor exposed corner is 2000mm high.

7) Exterior walls

Exterior wall 1: the construction practices of advanced exterior wall paint in creamy white can be found in L96J002——Exterior Wall 22.

Exterior wall 2: the construction practice of brick red advanced exterior wall paint can be found in L96J002——exterior wall 22.

8) Roofing

Roofing 1: The construction practices of dual-fortification insulation roof can be found in L96J002——House 27 and SBS membrane is used.

Roofing 2: The construction practices of dual-fortification can be found in L96J002——House 22.

9) Ceiling

The construction practices of topping out with latex paint at the plate bottom can be found in L96J002——Shed 9.

10) Doors and windows

The numbers, sub-grid sizes and practices of doors and windows can be found in J13.

The windows are 70-series steel windows with white frames and 5 thick white glass are adopted unless stated otherwise.

5. Others

1) The dimensions and elevations in construction drawings are expressed in the unit of meter and other parameters in construction drawings are expressed in the unit of millimeter unless stated otherwise.

2) The colors of the materials in the design should be negotiated jointly between the owner and designers according to color cards and provided samples before the construction.

3) The detailed practices of outdoor terrazzo brick steps and anti-slip strips can be found in L96J401——43-7.

4) The detailed practices of water apron can be found in L96J002——6-Apron 3 and its width is 1000.

5) The detailed practice of the glass blackboard can be found in L96J901——67-2 and the detailed practice of the lecture platform in the wooden classroom can be found in L96J901——72.

6) Stainless steel railing should be constructed with galvanized steel pipes.

7) The construction drawing shall be combined with that of other disciplines. It shall be denoted that all the non-mentioned items shall be implemented in accordance with the relevant national construction inspection rules. Any unknown issue shall be timely coordinated among related parties.

6. Drawing list

No	Drawing types	Series numbers	Drawing titles	Drawing sizes	Remarks
1	ACD	JS01	General construction instructions and drawing list	A2	
2	ACD	JS02	First floor plan	A2	
3	ACD	JS03	Second floor plan	A2	
4	ACD	JS04	Third floor plan	A2	
5	ACD	JS05	Fourth floor plan	A2	
6	ACD	JS06	Roof plan	A2	
7	ACD	JS07	South elevation	A2	
8	ACD	JS08	North elevation	A2	
9	ACD	JS09	East elevation, Western elevation	A2	
10	ACD	JS10	1-1 section	A2	
11	ACD	JS11	LT1 detail drawings	A2+	
12	ACD	JS12	Details of doors and windows	A2	

3.4 Construction Site Plan

1. Purposes of the construction site plan

The construction site plan shows the overall layout of a newly designed building. It displays the positions and orientation of a new building, as well as its relationships with surrounding environment (such as the outdoor field, terrain, landform, the original building, road layout, and landscape configuration).

A construction site plan shows property boundaries and means of access to the site as well as surrounding buildings related to the design. The construction site plan shall reflect the layout of a large area, including the new building, existing buildings and roads, landscape and other items. As for a construction project, the site plan also needs to show all the service connection pipes: drainage and sewer lines, water supply pipes, electrical and communications cables, and exterior lighting.

Different legends are adopted in the construction site plan to plot the peripheral contour lines of buildings (including new buildings, existing buildings, demolished buildings), surrounding roads, and green areas.

The construction site plan is the location basis for the newly designed building, the pipeline layout, and construction site.

2. General contents of the construction site plan

1) Titles and scales of drawings

Due to the large area of the construction site plan, small scales such as 1 : 500, 1 : 1000 and 1 : 2000 are usually used. A proper scale shall be chosen based on the area of the construction site as well as the size of the drawing sheet. The scale of topographic maps provided by the national bureau and relevant units is usually 1 : 500, so this scale is also usually adopted in the construction site plan.

2) Legends of the construction site plan

National standards (GB) have recommended different kinds of legends to indicate the general layout of the construction site, such as buildings (new buildings, expansion of existing buildings or renovation of existing buildings), floors, locations, roads, squares, outdoor fields and green areas. Moreover, other legends which are not included in GB may be used if necessary.

The Standard for General Layout Drawings (GB/T 50103—2010) has listed the legends for the construction site plan, such as legends of roads, railways, pipelines and landscapes. Some of them are provided in Table 3.4.1.

Legends for the construction site plan   Table 3.4.1

Names	Legends	Descriptions
New building		Bold lines are used to indicate the peripheral contour of a building (usually the locating axis or outer surface of the exterior wall at the elevation of ±0.000); "▲" indicates the main entrance of the building. The total storey number of the building is indicated with a number or several dots
Existing building	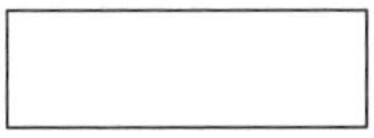	Thin lines
Planned building		Semi-bold dash lines
Demolishing building	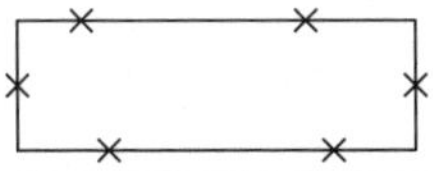	Thin lines
Fence and gate	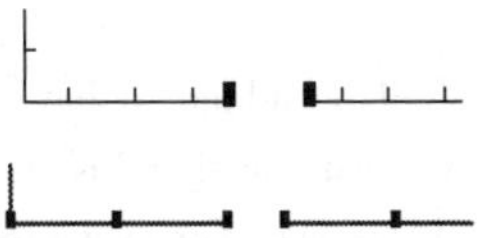	The upper figure shows the solid wall and the lower figure shows through walls. If only walls exist, the gate can be omitted
New road	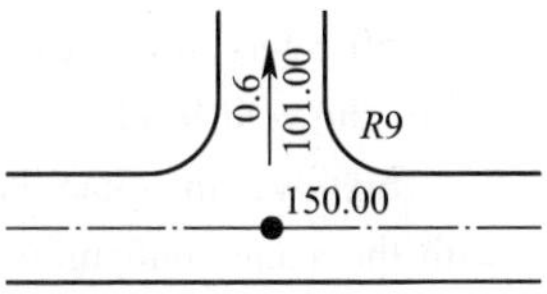	"$R9$" indicates the radius of turning circle of 9m; "150.00" refers to the elevation of road's middle control point; "0.6" refers to 0.6% longitudinal grade; "101.00" refers to the distance away from the gradient change point
Existing road	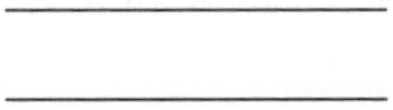	
Lawn	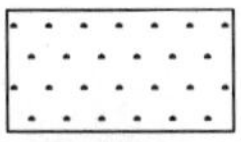	
Deciduous needle-leaved tree		

3) To determine the locations of the new, expanded or innovated building

The positions of the new, planned or rebuilt buildings is located according to the existing buildings or roads and relative distances are expressed in the unit of meter.

4) To show the absolute elevations of the first floor, outdoor terrace and roads

There are two types of elevation: namely, the absolute elevation and the relative elevation.

The absolute elevation is the reference elevation of a country or a region of uniform datum as the zero level. In China, the mean sea level of the Yellow Sea near Qingdao is taken as the absolute zero-level elevation. All the elevations in the construction site plan are the absolute elevations. Since there are many elevations to be indicated in the construction drawings of buildings, absolute elevations are only used in the construction site plan. In other drawings, relative elevations are usually used instead. In other words, the main floor height in the ground floor is taken as the zero-level of the relative elevation (indicated as "±0.000"). In the general instructions, the relations between the relative elevation and the absolute elevation shall be provided. The nearby benchmark (absolute elevation) is used to measure the absolute elevation of the floor in the ground floor.

In the **Unified National Standard for Building Drawings** (GB/T 50001—2017), the requirements for elevation symbols in housing construction drawing are introduced below:

Firstly, an isosceles right-angled triangle is used to represent the elevation symbol with thin lines, as shown in Figure 3.4.1(a). If there is not enough space to write the number, the form in Figure 3.4.1(b) may be used. In the elevation symbols shown in Figure 3.4.1(c) and (d), the height of the symbol is about 3mm.

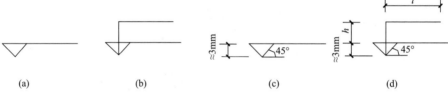

Figure 3.4.1 Elevation symbols
l- elevation; h- proper height

Secondly, the outdoor terrace elevation in the site plan and the first-floor plan shall be denoted with a solid triangle as the elevation symbol, as shown in Figure 3.4.2.

Thirdly, the tip of an elevation symbol should point to the indicated position. The tip may point upward or downward (Figure 3.4.3). The elevation figure shall be indicated on the right or left top of the elevation symbol.

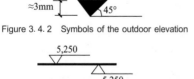

Figure 3.4.2 Symbols of the outdoor elevation

Figure 3.4.3 Direction of the elevation

Fourthly, the elevation figure shall be indicated in meters with three decimal places, but only two decimal places are required in construction site plan drawings.

Fifthly, the zero-level elevation shall be expressed in the form of ±0.000. The sign "−" shall be added to indicate negative elevations, whereas the sign "+" is not required for positive elevations. For example, 0.350 means the elevation of 350mm above the zero level and −0.350 refers to a negative elevation of 350mm below the zero level.

Sixthly, in some cases (different floors with the same configuration), it is necessary to mark several elevations at the same position in one single drawing, as shown in Figure 3.4.4.

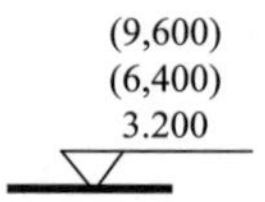

Figure 3.4.4 Several elevation figures at the same position

5) Wind rose and compass

A wind rose or compass is used to show the dominant wind direction. A compass may also be used to denote the orientation of the buildings. Sometimes, the wind rose can be omitted and only the compass is drawn.

A compass is usually drawn to point to the direction of north, as shown in Figure 3.4.5. The circle is plotted with thin lines and its diameter is 24mm in. A solid triangle is drawn at the center of the circle with a 3-mm wide tail at the south end, and a character of "N" at the north end. If a larger diameter is used to draw the compass circle, the width of the tail shall be 1/8 of the diameter. The symbol of the compass is prerequisite in the construction site plan and the first floor plan.

Figure 3.4.5　Compass

In some cases, the wind direction frequency of the area shall be denoted, then a wind rose may be drawn in the construction site plan to indicate the wind direction.

3. Case study

Figure 3.4.6 is the site plan drawing of the aforementioned laboratory building in a middle school. Some key items in Figure 3.4.6 are introduced as follows:

1) Drawing title, scale and orientation

The scale of the construction site plan is 1 : 500, as demonstrated under the drawing. The direction of the compass shows the orientation of the building.

It can be seen from the drawing that a four-storey laboratory building is going to be constructed in the northwest corner of the school. Since the scale of the drawing is small, only the contour shape, location and relations with surrounding environment of the new building is drawn and demonstrated. The terrain contour lines of the buildings are not required.

2) Surroundings

The west and north sides of the new laboratory building are surrounded by fences. The distance between the northern exterior wall and the enclosing fence is 14m. A five-storey teaching building is planned to be built in the east of the laboratory building. The distance between the laboratory and the teaching building is 18m. A five-storey teaching building in the south is 25m away from the laboratory building. There is a lawn along the northern wall of the existing teaching building, and a flower bed is sit between the newly built laboratory building and the existing teaching building.

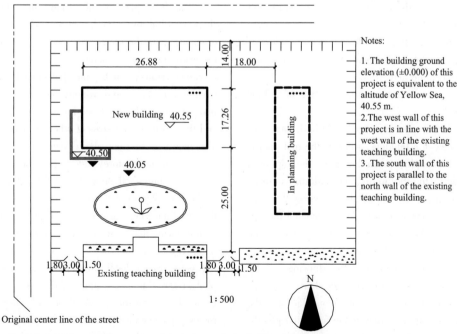

Figure 3.4.6　Construction site plan of the laboratory building in a middle school

There are two main entrances with the width of 3m on the east and west sides of the existing teaching building.

3) Information of the new building

The shape of the designed laboratory is plotted with bold lines. The length of the laboratory from the east to the west is 26.88m and that from the north to the south is 17.26m. The designed laboratory sits towards the south. The stories of the building are indicated with figures on the top right with small black dots (or numbers). It can be seen that the new laboratory is a 4-storey building.

The location of the new laboratory is determined by the positioning dimension. The western exterior wall of the laboratory building is in line with the western exterior wall of the existing teaching building. The distance between the exterior wall and the western enclosure fence is 6.3m. The southern exterior wall of the laboratory building is parallel to the northern western exterior wall of the existing teaching building with a distance of 25m.

The absolute elevation of the ground floor is 40.55m and the elevation of the outdoor ground is 40.05m, indicating that the altitude difference between indoor ground and outdoor ground is 500mm.

3.5 Floor Plan

A floor plan is the basic drawing among architectural drawings. It is the important basis of piping, wall construction, the installation of doors and windows. Other architectural drawings (such as the elevation, section, and detail drawings) are derived from the floor plan. In addition, the floor plan is the basis for the drawings of other disciplines (such as structure and equipment). Therefore, compared with other drawings, the floor plan contains more information, and the requirements for the drawing are more accurate and strict.

1. Formation process and title of a floor plan

A floor plan is actually the horizontal section view of a building, whereas the roof plan is the orthographic projection from the top of a building. In other words, the building is cut by an imaginary horizontal plan crossing the window and door openings. After removing the upper part, the lower part is projected onto the horizontal plan to form an orthographic projection drawing. Therefore, the drawing is also called the floor plan or plan for short.

Similar to maps, the floor plan has the downward view. However, the formation process of a plan is different from that of a conventional map since it is drawn at a particular vertical position (commonly at about 1-2 meters above the floor). Objects at this elevation level are shown by "sectioning"; objects below this level are not cut by the imaginary plan but still visible; objects above this level are invisible and can be omitted or drawn with dash lines.

The floor plan mainly shows the planar shape of a building, the layout and combination of each part in the horizontal direction (such as the main entrance, secondary entrances, rooms, corridors, stairs, toilets, and balconies). The positions of doors and window openings, the plan layout of the load-bearing wall (pillars) and the positions and sizes of other elements are also shown in the floor plan drawing.

As for multi-storey buildings, the corresponding name shall be added at the bottom of in the floor plan drawing of each storey, such as the first floor plan (also called the ground floor plan or basic plan), second floor plan, and roof plan. If the layouts of several floor plans are all the same except local differences, a common plan (referred to as the standard floor plan) can be used to represent these floors. As for the rooms with complex local layouts or more fixed equipment (such as kitchen and toilet), it is difficult to show the details of the rooms at the small scale of 1 : 100 or 1 : 200. Therefore, it is necessary to draw a local floor plan with a relative large scale in order to show more details.

2. First floor plan

The first floor of a building is a transportation hub between the interior and exterior spaces. As for a building with a basement, the ground floor is the transfer storey between underground and above-ground spaces. The entrances in different directions and rooms with different functions (such as lobby, reception room, and guardroom) may be arranged on the ground floor. Steps, ramps, and flower beds are arranged in the exterior space around the building. The design of the above-mentioned elements directly affects the interior functions and exterior appearances of the whole building. In addition, the plan of the ground floor usually determines the layouts of column grids (or load-bearing walls), sizes, locating axes and room layouts. Therefore, the first floor plan is the most basic drawing among all the architectural plans.

Figure 3.5.1 – Figure 3.5.5 show the examples of architectural floor plans of an office building. Due to different room layouts on each floor, the first floor plan, the second to fourth floor plans and roof plan are respectively shown. Firstly, taking the first floor plan in Figure 3.5.1 as an example, the contents and requirements of the architectural plans are illustrated.

1) Drawing title, scale and orientation

The title of the drawing is "First floor plan", which means the layout of the ground floor. The plan is a horizontal cross-section of the first floor obtained with the downward projection method after imaginary sectioning of this floor. The sectioning position is above the windowsill of the ground floor and below the stair platform between the 1st and 2nd floors. According to the size and complexity of the building as well as the size of drawing sheet, a scale of 1 : 100 is used. The orientation of the building is shown by the compass at a noticeable position and consistent with that in the construction site plan. The main entrance of the building is in the northwest corner and shall be emphasized in the first floor plan.

2) Numbering of vertical and horizontal locating axes

The office building is a frame structure building, in which the primary load-bearing structural system is the reinforced concrete frame. Therefore, the locating axes shall be determined based on the positions of the frame columns. The vertical locating axes are numbered from left to right (① to ⑦) and the horizontal locating axes are numbered from bottom to top (Ⓐ to Ⓓ).

The walls are made of 200mm thick aerated concrete blocks and used for spatial division. Since the walls are non-bearing components, additional locating axes are used to positioning them. For example, the numbers of (1/Ⓑ) and (1/⑥) indicate the first additional locating axes after axis Ⓑ and axis ⑥, respectively.

Based on the serial numbers of the locating axes and the distances between them, the position of each frame column is determined. For example, the distance between column ① and ② along axis Ⓓ is 7800mm.

3) Combination and partition of rooms and cross-section of walls and columns

The main entrance is located between axes ① and ② along the axis Ⓓ in the northwest corner of the building. The position of -0.020 can be reached after three steps from the outdoor terrace. One can enter the hall through the door M3935. The standard width of the lobby is 7800mm. Two reinforced concrete columns (800mm×800mm) are set on both sides of the main entrance. One secondary entrance with a door of M1824 can be found on the east of the building.

There is a staircase in the south of the lobby. Since the staircase plays an important role in fire evacuation, a fireproof door FM1521 is adopted. The upward and downward directions

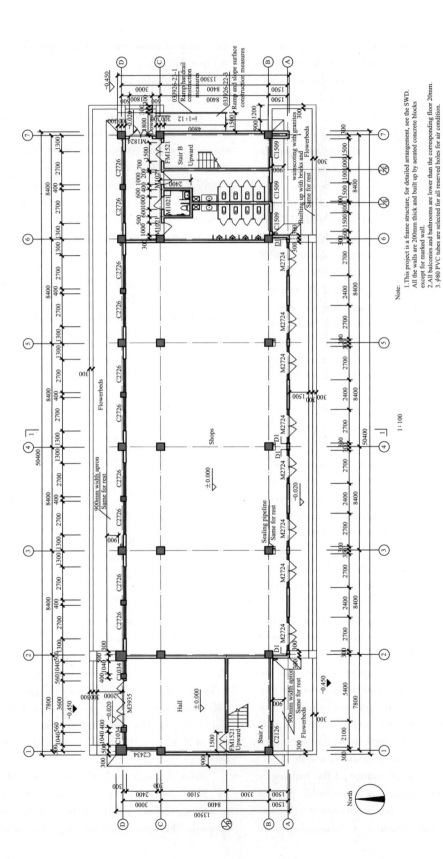

Figure 3.5.1 First floor plan

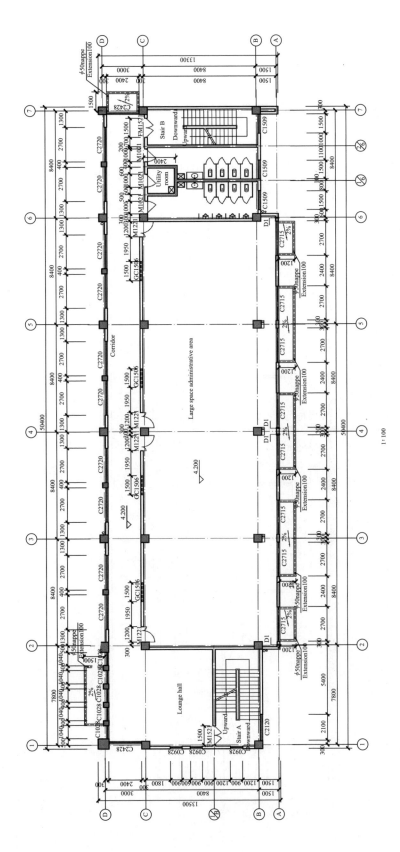

Figure 3.5.2 Second floor plan

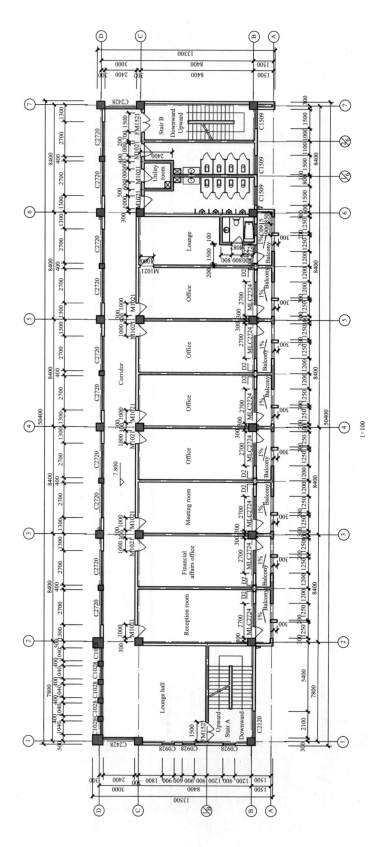

Figure 3.5.3 Third floor plan

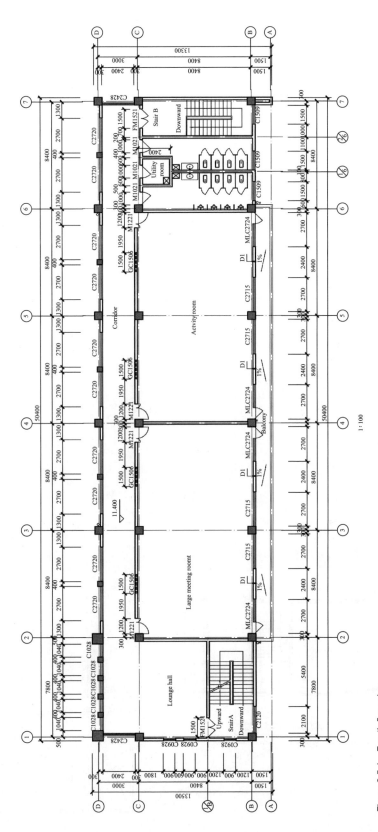

Figure 3.5.4 Fourth floor plan

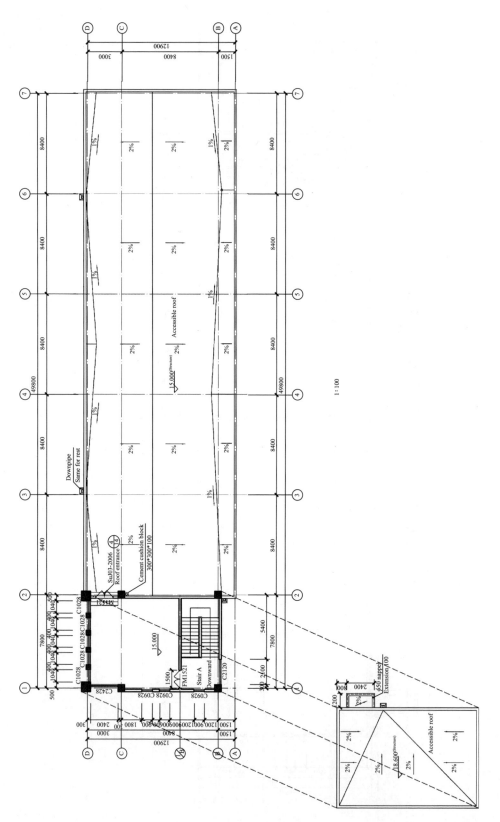

Figure 3.5.5 Roof plan

and main sizes of the stairs shall be denoted.

The office is a one-side corridor building. The distance between the two axes of the corridor denotes the width of the corridor, 3000mm and the depth of rooms is 8400mm. The depth and width meet the requirements of modulus coordination. Shops are arranged on the first floor. Offices, reception rooms, meeting rooms and activity rooms are arranged on the second to fourth floor. The toilets are adjacent to the staircase in the east. The floors of the toilets are 20mm lower than that in adjacent rooms. Two downpipes are set outside of the longitudinal walls in the north and south. There is an apron around the exterior wall with the width of 900mm.

4) Layout of doors and windows

The positions and numbers as well as the opening directions of doors and windows are shown in the plan. The types, numbers and positions of doors and windows are defined by the legends and serial numbers. The legends of doors and windows mentioned in related national standards are shown in Table 3.2.3. In the legend of a window, the two parallel thin lines represent the positions of the window frame and casements. In the legend of a door, the middle line shall be opened to form an angle of 90° or 45°and an arc may be added to show the opening direction.

The serial numbers of doors and windows are usually defined according to their materials, functions or dimensions. For example, a capital letter "M" is used to name a door; "MM" indicates a timber door; "GM" indicates a steel door; "SGM" indicates a plastic steel door; "LM" indicates an aluminum alloy door; "JM" indicates a rolling door; "FDM" indicates a security door; "FM" indicates a fireproof door. Similarly, other items and corresponding characters are provided as follows: window (C), wood window (MC), steel window (GC), aluminum alloy window (LC), plastic steel window (SGC), and fire window (LC).

Doors and windows with the same serial numbers have identical structures and sizes.

The forms and sizes of doors and windows can be found in the elevation drawing, section drawing, and the window and door tables. As for high windows on the inner wall of the corridor such as GC1506, dash lines may be used to indicate their positions above the sectioning plan.

In Figure 3.5.1, there are five types of doors in the first floor plan, numbered as M3935, M2724, M1824, M1021 and FM1521, respectively. The widths of these doors are respectively 3900mm, 2700mm, 1800mm, 1000mm and 1500mm and corresponding heights are respectively 3500mm, 2400mm, 2400mm, 2100mm and 2100mm. There are also five different types of windows, respectively numbered as C1034, C1509, C2126, C2434 and C2726 and their widths are respectively 1040mm, 1500mm, 2100mm, 2400mm and 2700mm.

The positions of other architectural components including steps, ramps, water aprons, open trenches, fire ladders, refuse channels, rain pipes, pools, platforms, cabinets, cupboards, and partitions should also be indicated in the floor plan. Figure 3.5.1 shows the design data of steps, water aprons, ramps, flower beds, etc.

5) Drawing lines

Different components in architectural drawings shall be drafted with different lines. The contours of the components that intersect with the sectioning plane shall be drafted with bold lines (b), such as the contour lines of walls and the columns. It should be noted that the contour lines of walls or columns do not include the depth of the plastering layer. The medium thick line ($0.7b$) is used to draw the contour lines of other components below the sectioning plane, such as windowsills, steps, water aprons, flower beds and benches. Dimensional lines and various symbols shall be drafted with thin lines. Locating axes shall be drafted with thin-dot-dash lines. Invisible components (such as the high windows above the sectioning position) and other invisible openings (spiracle, groove, trench and crane) shall be drafted with thin dash lines.

6) Material legends

The drawing methods of material legends under different scales may be simplified according to the ***Unified Standard of Housing Construction Drawing*** (GB/T 50001—2017).

In plan drawings with a scale larger than 1∶50, material legends should be drawn. In plan drawings with a scale between 1∶100 and 1∶200, material legends may be simplified (such as the solid square and reinforced concrete). In plan drawings with a scale of less than 1∶200, material legends can be omitted. In Figure 3.5.1, reinforced concrete columns are indicated with the simplified legend of solid square and the legend of brick masonry walls is omitted due to limited net space between contour lines.

7) Dimension annotations

The dimensions in architectural drawings must be denoted clearly. The dimensions in the plan drawing include the quantitative dimensions and positioning dimensions. The former refer to the physical dimension of architectural components, such as the thickness of walls, the size of column section, and the width of doors and windows. The positioning dimensions refer to the position dimension of components in the plan, such as the distance between a door and a locating axis and the distance between interior walls.

In architectural plan drawings, the dimensions include external dimensions and internal dimensions.

The external dimensions of the floor plan are generally divided into three sets of dimensions. The first set of dimensions include the dimensions of door and window opening in exterior walls and the dimensions of the wall between adjacent doors or windows (counted from the locating axis). In Figure 3.5.1, the width of windows in the longitudinal exterior wall are all 2700mm. The window width in the cross wall beyond axis ① is 2400mm and the width of doors are 2700mm or 1800mm. The second set of dimensions are the distance between locating axes, the distance between primary elements such as walls or the columns. The third set of dimension is the overall length between walls on opposite ends of the building. For example, the overall length of the building is 50400mm and the overall width is 13500mm.

In addition, in most drawings, internal dimensions shall be provided, such as the thickness of interior and exterior walls, the height and width of columns, the sizes of door and window openings in interior walls, the width of steps, water apron and stairs.

Besides the above-mentioned linear dimensions, the elevations shall be indicated in the first floor plan. The elevation dimensions usually include the elevations of the indoor and outdoor floors, flooring, top face of the step, balcony top face, stair landing, etc. In Figure 3.5.1, the relative elevation of the ground floor is set as the zero level, ±0.000. The outdoor elevation shall be indicated as −0.450 with a solid triangle. The elevation of the toilet is 20mm lower than that of the adjacent floor, indicated as −0.020. The step surface of the entrances is indicated as −0.020, indicating that the elevation at this point is 20mm lower than the lobby floor.

8) Sectioning, symmetry and indexing symbols

In first floor plan drawing, the sectioning symbols at the sectioning position for the section drawing should also be indicated.

In the plan drawings of symmetrical buildings, a symmetry symbol can be used as the symmetrical boundary of the building and it is only necessary to draw half of the building plan. In this way, the drawing sheet and time can be saved.

In addition, indexing symbols should be used to indicate the indexed drawing if necessary. For instance, the indexing symbol of 03J926−23−1 is indicated at the handrail of the ramp in Figure 3.5.1, which means that the practice shall refer to Figure 1 in Page 23 of ***the Barrier-Free Design Atlas of Buildings*** (03J926).

9) Others

The functions and serial numbers of rooms should be indicated in floor plan drawings.

If the detailed sizes or construction methods of some parts of a building are not indicated in the drawing, explanatory notes shall be provided.

As for the long floor plan of buildings, the partitioning drawing method may be adopted. In addition, the schematic diagram of the combination shall be given in the ground floor of each partition and the partition number should be indicated clearly.

Although the types and sizes of various doors and windows can be found in the floor plan, the table of doors and windows (Table 3.5.1) shall be provided to facilitate civil construction and manufacturing. The door and window table of small- and medium-sized buildings are usually included in architectural drawings. According to Deep Regulations, the door and window table shall contain categories, design number, opening size, frame quantity standard atlas number, remarks, etc. The remark column should contain the features of doors and windows (such as the opening direction and functions), materials or accessories (such as the screen door) and additional contents (such as the presence/absence of door sill and the distance between window top and beam bottom).

Table of doors and windows   Table 3.5.1

Types	Design No.	Opening sizes (mm)		Set	Remarks
		Width	Height		
Doors	FM1521	1500	2100	9	Class-B fire door
	M2724	2700	2400	8	Finished wood doors
	M3935	3920	3500	1	Safety glass door
Windows	C1509	1500	900	12	Double-glass screen window
	C2434	2400	3350	1	Double-glass screen window
	C2726	2700	2600	10	Double-glass screen window

3. Other floor plans

The contents and requirements of the first floor plan have been introduced in 3.2. Taking the office building as an example, the contents and requirements of other floor plans, roof plan and local plan are briefly introduced below.

1) Mid-floor plans

Figure 3.5.2 – Figure 3.5.4 respectively show the 2nd floor plan, 3rd floor plan and 4th floor plan of the office building. The formation process of these drawings is the same as that of the first floor plan. The mid-floor plan is the horizontal section of the downward projection near the middle height of this floor. The sectioning position is above the windowsill and below the stair platform to the upper storey. The contents of the floor plan are basically the same with the first floor plan. However, the outdoor components and equipment, such as flower beds, outdoor steps, open trench, and water apron, as well as the compass and sectioning symbols included in the first floor plan drawing need not to be drawn in the mid-floor plans.

It should be noted that the drafting method of the stair in mid-floor plan is different from that of the first floor plan. In the first floor plan, only half run of the upward stair steps are visible. In the mid-floor plans, both the upward stair steps and the downward stair steps are visible, so the full run of the stairs shall be drawn. Since the stairs are slopes, an inclined broken line is used to show the imaginary sectioning. In the top floor or roof floor plan (in case of accessible roof), only the downward stair steps are visible in the projection and the broken line is not needed. The elevations of the resting platform of the stairs shall be indicated in the floor plan drawing.

In the 2nd floor plan (Figure 3.5.2), the rain cover and drainage conditions are indicated. The drainage slope is 2%. The drainage port is arranged and the finished water tongue is used for drainage. Although the rain cover in 2nd floor plan is also visible in the 3rd and above floor plans, it may be omitted in these drawings for simplification.

2) Roof plan

The roof plan shows the view from the top of the roof. Figure 3.5.5 is the roof plan of the office building. The roof plan is relatively simple, thus a larger scale (such as 1 : 200) can be used. The roof plan usually contains the parapet, cornice, gutter, gradient, exposure, gutter inlet, ridge, deformation joint, staircase, water tank, elevator room, sunroof wind deflector, roof manhole, access ladder, outdoor fire ladder and other components.

In Figure 3.5.5, the roof is a south-north double drainage system and the drainage slope is 2%. There is a drainage ditch inside the parapet in the south and north. The drainage slope of the ditch is 1%. There is a drainage port in the ditch connected with external downpipes.

3) Local plans

As for the rooms with special equipment or complex layouts, it is difficult to reveal local details in the rooms clearly in the plan with a scale of 1 : 100. Therefore, the local plans with a larger scale are required for accurate construction. The local plans usually include toilets, rooms with special equipment, ramps of garage, etc. Figure 3.5.6 is the local plan of toilets with a scale of 1 : 50. The dimensions and positioning dimensions of some sanitary fixtures are shown in Figure 3.5.6.

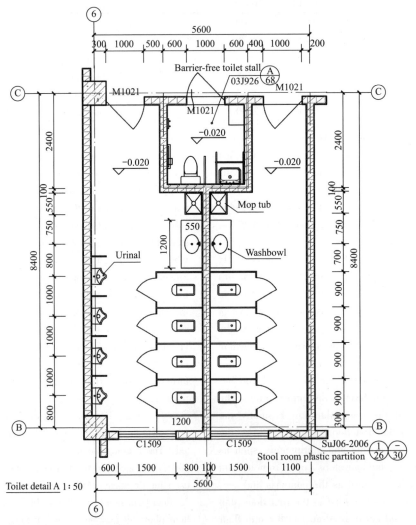

Figure 3.5.6 Detail drawing of toilets

4) Drawing steps of floor plans

Before drawing, it is necessary to firstly determine a proper scale according to the size and complexity of the building. Then, the required positions of sizes, symbols and the illustrations can be estimated. Next, a proper drawing sheet can be chosen and the frame and title bar are drawn to leave an enough drawing area.

Taking the first floor plan of the office building as an example, the drawing steps are given as follows (Figure 3.5.7):

(1) To draw the locating axes of walls (columns) (Figure 3.5.7a).

(2) To draw the contour of components, such as walls, column sections, and positions of doors and windows (Figure 3.5.7b).

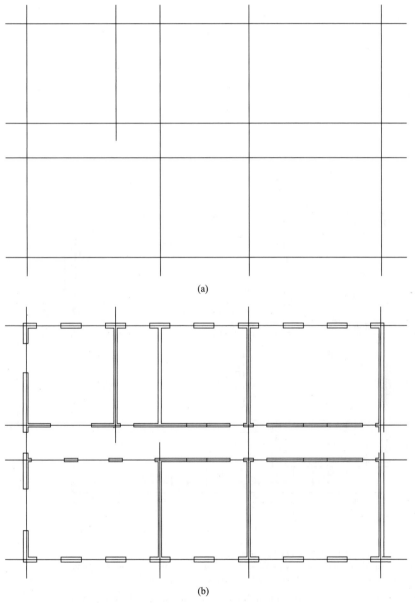

Figure 3.5.7 Drawing steps of an architectural plan (one)
(a) Draw steps of locating axes of walls (columns);
(b) Drawing steps of contour lines of components and the positions of door and window openings;

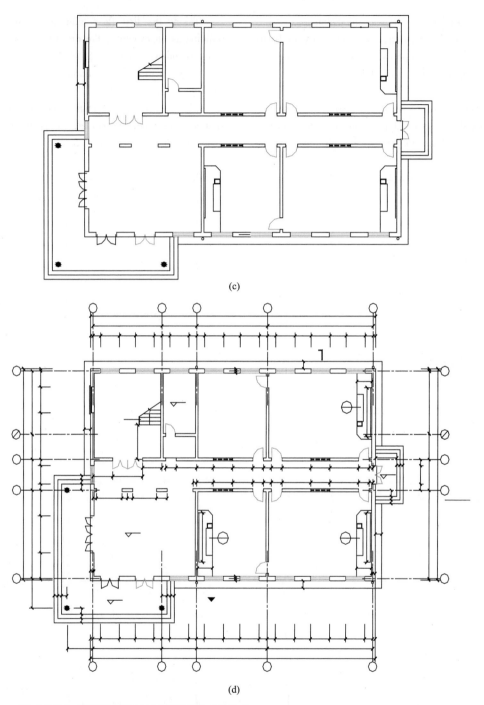

Figure 3.5.7 Drawing steps of an architectural plan (two)
(c) Drawing steps of the details, such as doors and windows, steps, stairs, water apron, downpipe, platform and desks;
(d) Drawing steps of dimension lines, elevations, and sectioning and indexing symbols

(3) To draw the details or contour lines, such as doors and windows, steps, stairs, water apron, downpipe, platform, sanitary fixtures and desks (Figure 3.5.7c).

(4) To draw the dimension lines and various symbols, such as elevation symbols, sectioning

symbols and indexing symbols (Figure 3.5.7d).

The extra auxiliary drawing lines shall be removed after checking the drawing. Then the lines should be deepen according to the requirements. The sizes and elevation, indexing and sectioning symbols should be indicated.

In addition, the names of rooms, the serial numbers of doors and windows, the title and scale of the drawing, and necessary explanatory notes should be added.

3.6 Elevation Drawings

1. Formation and name of an elevation drawing

A building elevation drawing is a horizontal orthographic projection of the building onto a vertical plan, which is normally parallel to one side of the building. Figure 3.6.1—Figure 3.6.4 show the four elevation drawings of the above-mentioned office building from different directions.

The building elevation drawing shows the exterior views of a building, including shapes, positions and opening directions of doors and windows in exterior walls as well the positions, dimensions and methods of other finished elements (such as skirting, rain pipe, lead line, windowsill, cornice, roof tank, balcony, and rain cover).

The building elevation drawing can be named with several methods. The building with the locating axes can be named with the axis number (such as ① ~ ⑦ elevation drawing and Ⓐ ~ Ⓓ elevation drawing). The elevation drawing without locating axis can be named with the orientation, such as the east elevation drawing, the south elevation drawing, the west elevation drawing and the north elevation drawing. Sometimes, the elevation drawings are named with the main entrance, such as the front elevation drawing, the back elevation drawing, the left side elevation drawing and the right side elevation drawing.

2. Contents and requirements of elevations

Taking the north elevation drawing of the office building in Figure 3.6.1 as an example, the contents and requirements of the elevation drawing are explained as follows:

1) Drawing title, scale and locating axes

The elevation drawing is the ⑦ ~ ① elevation drawing, indicating the elevation facing north and the elevation with the main entrances. The elevation drawing can be named south elevation drawing or main elevation drawing. The scale of the elevation drawing is 1 : 100, which is usually the same as that of the architectural plan drawing.

To facilitate the comparison between the elevation drawing and the plan drawing, the elevation drawing usually includes two-end locating axes and numbers.

2) Full view of exterior walls above the outdoor terrace

The elevation drawing contains the types, positions and opening directions of doors and windows in exterior walls, shape of roof, as well as positions and shapes of steps, flower beds, rain covers, windowsills, balconies, rain pipes, fountains, finished exterior walls, and various architraves.

In the figure, the outer contour lines covers the overall length and height of the building composed of four floors. The contour lines and types of doors and windows are drawn according to actual situations. There is a gate at the main entrance, and the rain cover is located at the first floor. The forms of doors and windows shall be shown according to the requirements in National Standards.

According to the requirements in National Standards, as for the same doors and windows, balcony, eaves methods, construction practice could be highlighted, and other parts, it is only necessary to draw contour lines.

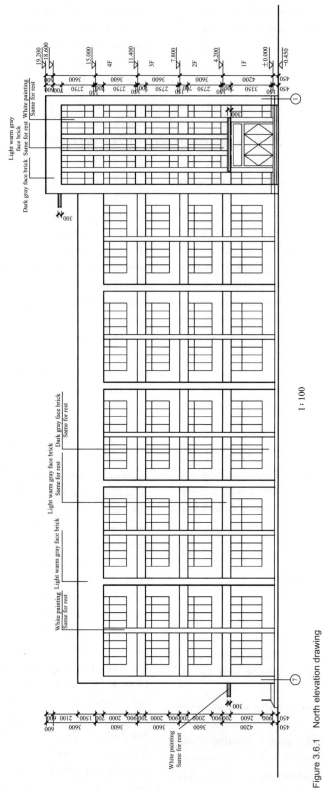

Figure 3.6.1 North elevation drawing

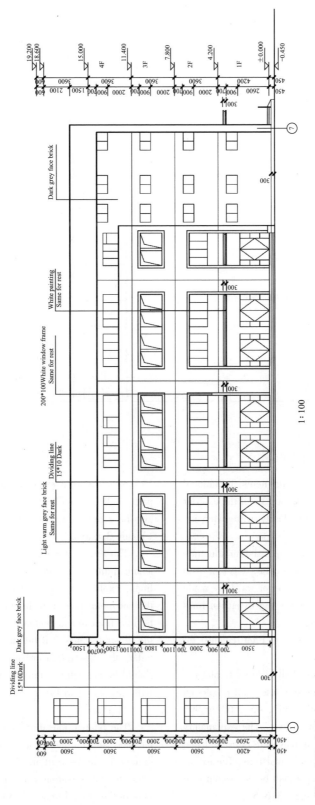

Figure 3.6.2 South elevation drawing

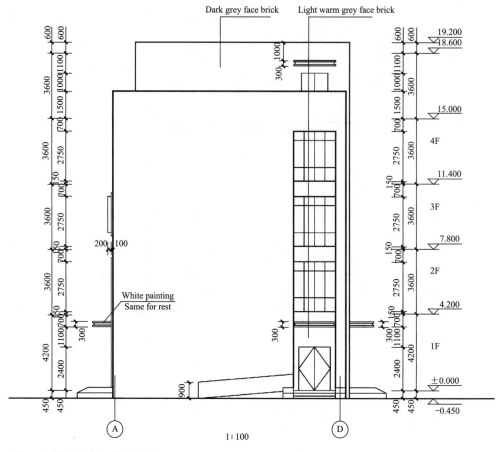

Figure 3.6.3　East elevation drawing

According to the requirements in National Standards, the finishing practices of the exterior wall shall be indicated clearly with extension lines.

The extension lines are used to illustrate each part's materials and colors together with texts. In the figure, the exterior walls are made of the dark grey face bricks and extension lines indicate white painting.

The followings are the requirements for "extension lines":

Firstly, thin lines shall be used to plot extension lines. An extension line shall be plotted to form the angle of 0°, 30°, 45°, 60° or 90° with a horizontal straight line. The explanatory notes shall be arranged above the horizontal line or at the ends of the horizontal line. The extension lines in detail views shall be connected with the horizontal line (Figure 3.6.5).

Secondly, the extension lines of the same parts shall be parallel to each other. They can also be arranged in the pattern of radial lines (Figure 3.6.6).

Thirdly, as for the extension lines shared among multi-layer structures or multi-layer pipelines, explanatory notes shall be added above the horizontal line or at the ends of the horizontal line from top to bottom and consistent with the illustrated layer. If the order is a horizontal order, the description order from top to bottom should be consistent with the layers from left to right, as shown in Figure 3.6.7.

3) Dimensions

The relative elevation is mainly used in dimension annotations. Sometimes, the vertical dimension can be written. The relative elevation refers to the elevation compared to the indoor floor of the ground floor (the elevation is zero).

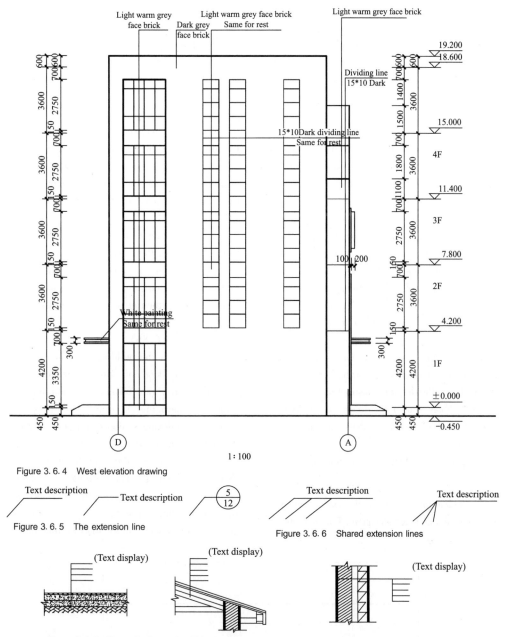

Figure 3.6.4 West elevation drawing

Figure 3.6.5 The extension line

Figure 3.6.6 Shared extension lines

Figure 3.6.7 Extension lines of multi-layer structures

Some elevations shall be indicated, including the indoor and outdoor elevations, and the upper and lower edges of door and window openings, parapet top, rain covers, balcony bottom, etc. They shall be arranged along a vertical straight line clearly and orderly. The construction elevation shall be distinguished from the structural elevation. The door or window openings do not include the plaster layers. The construction elevation is usually used to indicate the top surface of finished elements (such as parapet top surface and top surfaces of balcony railings), whereas the structural elevation is used to indicate the bottom surface (such as the bottom surface of the balcony and bottom surface of rain cover).

The positions of extension lines in vertical

dimensions should be consistent with the elevations. The dimension figures are the elevation differences.

If necessary, some local dimensions which are not included in detail drawings can be indicated in order to add sizes and positioning dimensions.

4) Lines

The facade of a building is usually drawn with thick lines (b) to make the elevation more clear. The outdoor floor is drawn with extra thick line ($1.4b$). Middle thick lines are used to draw main contours of door and window openings, rain covers, cornices, balconies and steps. Thin lines ($0.5b$) are used to draw the secondary contours of elements, such as doors and windows, frame line of wall, rain pipes, extension lines, dimension lines and elevation symbols. The lines ($0.25b$) are used to draw details.

5) Other symbols

As for simple symmetric buildings or components, it is necessary to draw half of the elevation without affecting the construction. The symmetric symbols should be drawn at the axis of symmetry. The indexing symbols should be used if necessary.

Figure 3.6.2–Figure 3.6.4 show ① ~ ⑦ elevation, Ⓐ ~ Ⓓ elevation and Ⓓ ~ Ⓐ elevation, which are respectively the south elevation, west elevation and east elevation. The contents and requirements of these elevation drawings are the same as those of the north elevation drawing.

3. Drawing Steps of Elevation Drawings

Similar to the steps of the architectural drawing, it is also necessary to firstly choose a proper scale and a drawing sheet before starting the elevation drawing. Then, the lines should be drawn and deepened. The drawing steps (Figure 3.6.8) are introduced as follows:

1) To draw the terrestrial horizontal lines, locating axes, floor line, cornice height contour lines and outer contour lines of the building (Figure 3.6.8a).

2) According to the corresponding architectural drawing, the positions of doors and window openings in the corresponding floor are drawn. The contour lines of door and window openings can be then obtained by measuring the vertical dimensions from ground lines and floor lines (Figure 3.6.8b).

3) To draw the building elements and details (such as door and window legends, windowsills, steps, rain covers, division lines of walls, and downpipes), dimension lines, elevation and indexing symbols and specifications of exterior walls (Figure 3.6.8c).

After checking, the extra auxiliary lines should be wiped off. The drawing lines should be deepened according to requirements of construction drawings. The drawing title, scale and necessary explanatory notes shall be included.

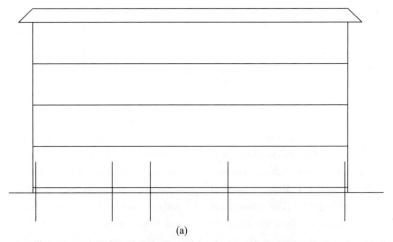

(a)

Figure 3.6.8 Drawing steps of elevation (one)

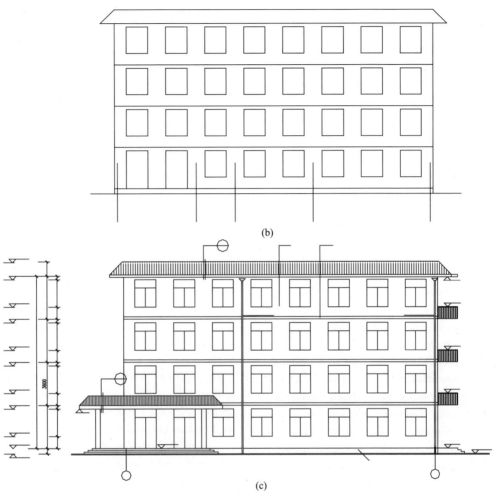

Figure 3.6.8 Drawing steps of elevation (two)

3.7 Architectural Section

1. Formation processes, names and purposes of section drawings

A section drawing of a building is the vertical section view of a building. A vertical sectioning plan parallel to the wall is assumed to cut the building from the roof to the foundation. Then, the orthographic projection obtained by projecting the remained parts onto the projection plan parallel to the sectioning plan is the section drawing, section for short.

The sectioning position shall be determined based on drawing purposes or design depth in order to clearly show interior conditions of a building, including free height of each room and vertical connections, height and materials of each part. The sectioning position should be able to reflect the full view and structural features. The general sectioning position goes through doors, windows and stairs so that the positions and structures of door and window openings and stairs can be displayed. The broken lines can be used to denote the basic parts of the section. The contents can be found in the basic drawing in structural construction drawings.

A sectioning symbol is composed of sectioning lines and projection direction lines. The above-mentioned sectioning symbols should be drawn with bold lines. The length of the sectioning line should be 6–10mm. The projection direction line shall be perpendicular to the sectioning line with the length of 4–6mm. The length shall be less than the sectioning line, as shown in Figure 3.7.1. During the drawing, the sectioning symbols shall not contact with other drawing lines.

The sectioning symbols can be numbered with Arabic numerals, Roman numbers or Latin letters. They should be arranged continuously from left to right and from bottom to top on the ends of the sectioning direction lines.

The number of section drawings depends on the building's complexity and actual construction requirements. A section drawing can be named with the number of corresponding sectioning symbol. For instance, the section with the sectioning symbol of 1 is named as 1–1 section.

One sectioning plan is usually used for section drawing. If one sectioning plan can not meet the requirements, the plan can be bent to draw the section drawing. The sectioning symbol is drawn in the first floor plan. The bent sectioning lines should be indicated and numbered outside of the turning corner.

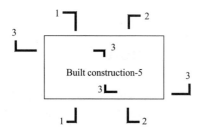

Figure 3.7.1 Sectioning symbol

Sectioning symbols of cross-sections (Figure 3.7.2) are also introduced in related regulations. The sectioning symbols of cross-sections can only be expressed with sectioning lines and drawn with bold lines with the length of 6–10mm. They should be arranged and numbered continuously with Arabic numerals. The number should be indicated on one side of the position line.

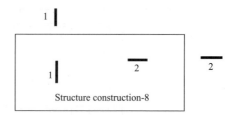

Figure 3.7.2 Sectioning symbol of cross-sections

If the section drawing or cross-section drawing is not in the same drawing with the sectioned drawing, the drawing number can be indicated on the other side of the sectioning line or stated in the drawing.

2. Contents and requirements of the section drawing

Figure 3.7.3 is the No. 1–1 section of the building.

1) Drawing title, scale and locating axes

The drawing title is 1–1 section. The proper scale is 1∶50, 1∶100 or 1∶200 depending on the dimensions and complexity of a building (Table 3.2.1). The scale is usually the same as that in the plan and elevation drawings. The adopted scale is 1∶100. When a larger scale is used to draw the section drawing, material legends shall be added in the cross sections of components and accessories.

In a section drawing, the locating axes and their gap of sectioning walls or columns shall be drawn, such as (A)-(D) locating axes in 1–1 section. When drawing and reading the section, the section should be compared with the plan. The relative positions of locating axes in the section should be consistent with the projection obtained by the plan projection along the sectioning direction.

2) Sectioned components and non-sectioned visible components.

In a section drawing, sectioned components and non-sectioned visible components shall be drawn.

Sectioned components include the positions, shapes and legends of indoor and outdoor floor (including steps, open trenches, and water aprons), floor (including ceiling), roof (including thermal insulation and waterproof

Chapter 3 Architectural Construction Drawings

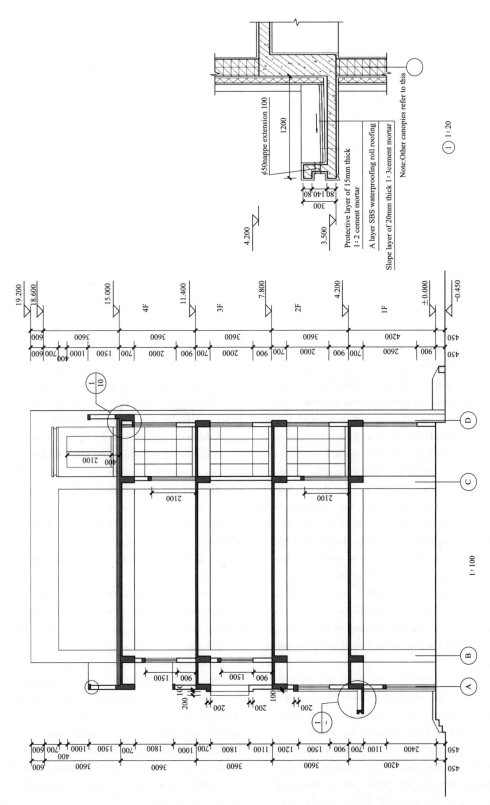

Figure 3.7.3 Section drawing of a building

layer and ceiling), interior and exterior walls, doors, windows (including lintel, ring beam, damp-proof course, parapet wall and ceiling), various bearing beams, link beams, stairs, stair platforms, rain covers, balconies, ducts, and water tanks. Figure 3.7.3 shows the legends of indoor and outdoor floors, 2nd to 4th floor, roof, parapets, cornices, exterior walls at (A) and (D) axes, interior walls at (B) and (C) axes, windows in exterior walls and doors in interior walls.

Visible parts include positions and shapes of walls, beams, columns, balconies, rain covers, doors, windows, skirting, plinth, steps, rain pipes, and staircases. In Figure 3.7.3, the visible parts along the sectioning direction include windows in the west wall of the corridor, top of the western parapet from the roof and the staircase protruding from the roof.

3) Drawing lines

Different lines are used in a section drawing. The lines of $1.4b$ are used to draw the indoor and outdoor terrestrial horizontal lines.

In the section drawing with a scale of 1 : 100, two bold lines (b) can be used to indicate the overall thickness of the structural layer and surface layer, such as sectioned rooms, corridors, stairs, and platforms. In the section drawing with a scale of 1 : 50, the surface layer should be expressed with two bold lines together with a line of $0.25b$ on the top. In general, the depth of plastering on the base is not given. The line of b is used to draw the sectioned contour lines of walls. In the section drawing with a scale of 1 : 100, the depth of plastering is not included. In the section drawing with a scale of 1 : 50, thin lines are used to indicate the depth of plastering.

Visible contour lines are drawn with the lines of $0.7b$, such as the door and window openings, stairs, handrails, parapet top, interior and exterior wall contour lines, skirting, and plinth.

The lines of $0.5b$ are adopted to draw dimension lines, size indication, elevations, doors, windows, frames, rain pipes, and exterior wall frame lines. The lines of $0.25b$ are used to draw legend filling lines and pattern lines.

4) Dimensions

Some necessary dimensions should be given in a section drawing, including dimensions and elevations.

External dimensions usually refer to three sets of vertical dimensions. The innermost dimensions refer to the heights of door and window openings and their gap. The middle dimensions refer to the floor heights including the heights from the floor of the ground floor to the 2nd floor, the height from each floor to the upper layer, the height from the top floor to the roof of cornice, the difference between indoor and outdoor floors, and the height from cornice to the parapet top. The outermost dimensions refer to the overall height above the outdoor floor.

The interior dimensions include the dimensions of door and window openings in the interior walls, handrails, eaves and rain covers.

The elevations shall be indicated, including the elevations of ground, floors, stair landing, balconies, cornices, rain covers, beam bottom, etc. The elevations of the floor, stairs, platforms and balcony top refer to the elevations of finished surfaces, which are also building elevations or elevations of the plaster layer. The elevations of rain cover bottom and beam bottom refer to the elevations of rough surface, which are also the structural elevations or elevations without the plaster layer.

In a section drawing, besides the dimensions and elevations in the vertical direction, the dimensions in the horizontal direction shall be indicated properly, such as the dimensions between locating axes in the sectioning direction.

5) Others

Indexing symbols are required for detail drawings. The extension lines and textual specifications can be used to indicate the construction materials and methods of ground, floor, roof, and interior walls according to the multi-layer construction order. In the figure, the above-mentioned methods have been elaborated clearly in building instructions, so they are omitted in the section drawing.

3. Drawing steps of section drawings

The drawing steps of a section drawing are basically the same with those of plan and elevation drawings. Firstly, it is necessary to choose a proper scale and a drawing sheet . Then, the lines are drawn and deepened. The general steps are introduced as follows (Figure 3. 7. 4):

1) To draw the locating lines, including horizontal locating axes, vertical indoor and outdoor terrestrial horizontal lines, floor lines, roof lines, and parapet top lines (Figure 3. 7. 4a).

2) To draw the positions of wall contour lines, ground lines, floor lines, roof lines, door and window openings as well as main components (Figure 3. 7. 4b).

3) To draw visible parts and details, such as rain covers, balcony, steps, door and window frames, and eaves (Figure 3. 7. 4c).

4) To mark dimensions, symbols, numbers and specifications (Figure 3. 7. 4d).

After checking, the extra auxiliary drawing lines shall be wiped off. The lines shall be deepened according to the requirements of the section drawing. The dimensions, elevations and specifications should also be given.

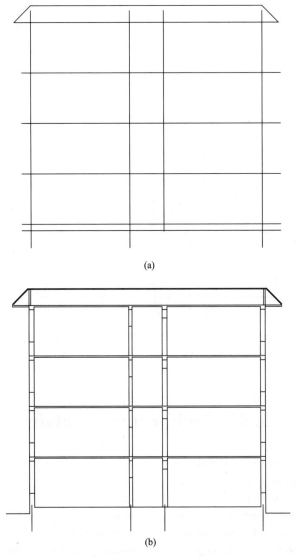

Figure 3. 7. 4　Drawing steps of a section drawing (one)

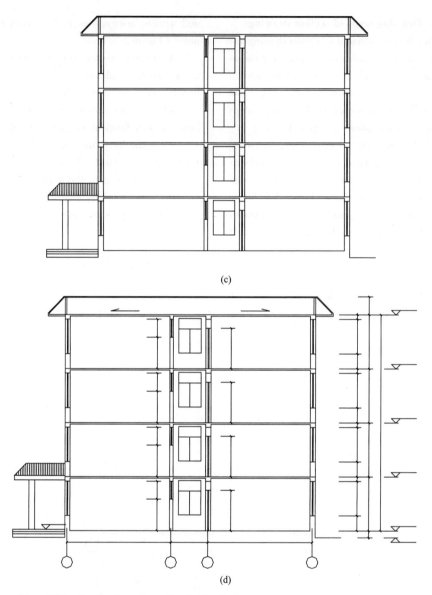

Figure 3.7.4 Drawing steps of a section drawing (two)

3.8 Architectural Details

The plan, elevation and section drawings are the basic drawings of architectural drawings with relatively small scales (1 : 100 and 1 : 200). Some components (such as doors, windows, stairs, balconies, and various decorative components), some section node drawings (such as cornices, windowsills, open trenches, floor or roof top) or detailed structures (including types, levels, methods, materials, or detailed dimensions) can not be expressed clearly. To meet the requirements of construction, it is necessary to provide the drawings with relatively large scales (such as 1 : 50, 1 : 20, and 1 : 10), which are called architectural details or detail drawings. The detail drawings for architectural components are called architectural details, whereas the detail drawings for structural components are called structural details.

Detail drawings are characterized by large scales, clear graphical representation, complete dimensions and detailed specifications. Therefore, detail drawings are supplements for plan and elevation drawings and indispensable in a complete set of construction drawings. Detail drawings are also one of the basis for the completion of design ideas during the construction. As for architectural components and section nodes on the standard atlas or general drawing, only the name of the atlas and the page number of the drawing are required to index the drawing.

The architectural details usually include detail drawings of rooms with special equipment (such as toilet and laboratory). The detail drawings are required to indicate positions, sizes and shapes of fixed equipment, embedded parts, grooves, and arrangement and constructional details (such as walls and interior and exterior surfaces, stairs, handrails, steps, ceilings, fire ladders and trenches).

Some detail drawings of an office building, including details of windows, doors, and stairs are introduced below to elaborate the contents and graphic methods. The reading methods and drawing steps of an architectural detail are also introduced.

1. Detail drawings of doors and windows

There are various doors and windows in the building. The detail drawing of doors and windows is used to indicate the dimensions, types, opening methods of doors and windows. At present, doors and windows are generally designed or selected according to standard atlas or general atlas. Doors and windows are usually produced by the factory. Then finished doors and windows are transported to the site and installed. If the standard atlas is available, only the standard atlas, general atlas or the types of doors and windows are indexed in the detail drawing. The details of doors and windows are unnecessary. Sometimes, it is only necessary to mark the elevations indicating boundary dimensions and opening directions of doors and windows and other details can be found in the atlas. If the standard atlas is not available, the detail drawings of doors and windows shall be drawn.

Figure 3.8.1 is the door and window detail drawing of an office building, where M1221 is the finished solid wood door; M1824, M2724 and M3935 are safety glass doors; MLC1524 and MLC2724 are double-glass doors and windows; C0915 ~ C2434 are double-glass windows with screenings; GC1506 are all double-glass windows.

The detail drawings of the facade of doors and windows are external views indicating dimensions and opening directions of doors and windows. The opening directions of doors and

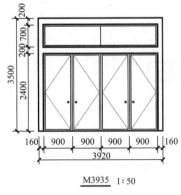
M3935 1 : 50

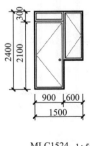
MLC1524 1 : 50

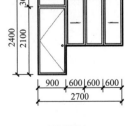
MLC2724 1 : 50

Figure 3.8.1 Detail drawing of doors and windows (one)

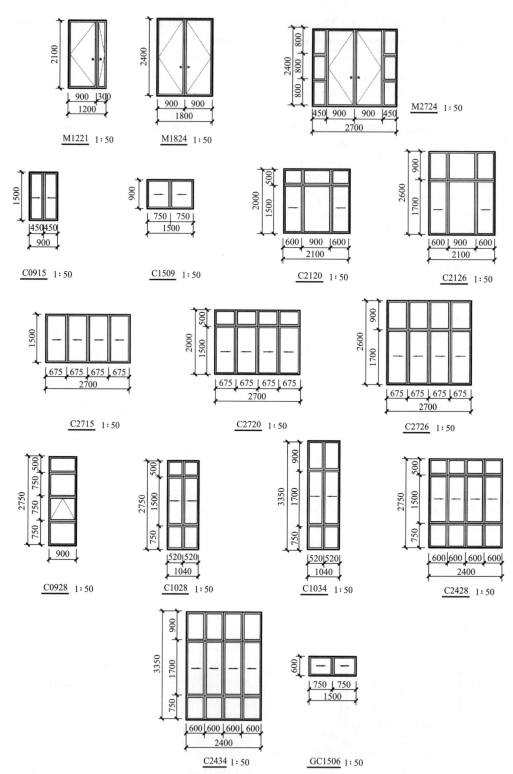

Figure 3.8.1　Detail drawing of doors and windows (two)

windows can be indicated with fine oblique lines. The side of the angle of the opening direction line indicates the installation position of the hinge. The solid line indicates the outward opening door, whereas the dotted line indicates the inward opening door. Arrows are used in sliding doors and windows to show the opening direction. The facade forms of doors and windows should be drawn based on actual conditions. Limited by materials, transportation and installation conditions, the facade division should consider the maximum limits of open doors and fixed doors. The dimensions in the elevation drawings of doors and windows include opening dimensions, production dimensions, installation dimensions and frame dimensions.

2. Stair details

Currently, the most widely used manufacturing methods of reinforced concrete stairs include the in-situ casting, prefabrication and the combination of in-situ casting and prefabrication. According to the plan shape of stairs, they can be classified into straight stairs, double-run stairs, multi-run stairs, shearing stairs, spiral stairs and etc. According to the structures, they can be classified into beam-supporting stairs and plate-supporting stairs.

The stair is composed of bench, stair landing, handrail, and stairwell. The detail drawing of stairs is mainly used to show the plate form, structure, dimensions and decoration elements of the stair.

The detail drawings of stairs include architectural details and structural details of stairs. Normally, they are drawn separately. As for a simple stair, architectural details and structural details can be drawn in the same drawing sheet and included in architectural drawings or structural drawings. The detail drawings of stairs are generally composed of plan (or local), section (or local) and node detail drawings.

Taking the stair detail drawings of the above-mentioned office building as an example, the contents and requirements of stair plan and section drawings are introduced below.

1) Stair plan drawing

Figure 3.8.2 is the architectural plan drawing of the stair A with a scale of 1 : 50.

The stair plan is also a horizontal orthographic projection. It is the downward orthographic projection obtained by sectioning the staircase of each floor with an imaginary horizontal sectioning plan. It contain the stair plans of each floor and roof. It should be noted that the sectioning position is below the resting platform except the top storey above the handrail. Generally, the stair plan should be drawn for each floor. However, if the types and structures of stairs on different floors are totally the same, it is only necessary to draw the first floor plan, the middle floor plan and the roof floor plan. In Figure 3.8.2, the stair plans and structures of the second, third and fourth floors are totally the same, thus only a standard stair plan is drawn and a series of elevations are added to show the omitted parts.

As shown in the drawing, Stair A is a two-run stair with a decline angle of 30°.

In the first floor plan, only the part of the stair from the first upward bench to the broken line is drawn and a broken line symbol is drawn in the sectioning plan. A long arrow is drawn in the middle of the projection with the letter of "upward" to indicate the moving direction from 1st floor to the 2nd floor.

In the 2nd and 3rd floor plans, since the sectioning plan is blow the middle platform, only half of the upward bench is visible by the projection method and drawn blow the broken line with the letter "upward". The downward steps blow the current floor are still visible and thus more "downward" steps are drawn.

In the top floor stair plan, the position of the sectioning plan is above the banisters, thus only downward stairs are drawn. The number "22" along with the "downward" arrow indicates that one can reach the lower floor by walking 22 steps. Handrails or breast boards shall be set on the suspension side of the top floor to ensure the safety. Therefore, the projections of handrails or breast boards should be drawn in the top floor stair plan.

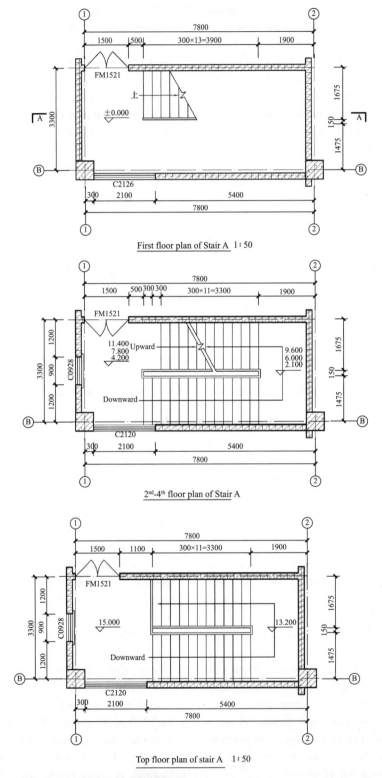

Figure 3.8.2　Section drawing of Stair A

The dimensions in the stair plan usually includes the positions of staircases indicated by axis numbers, length of the staircase, the number of runs, width of each run, number of steps, width of steps, dimensions of landing board, etc. In addition, the elevations of each floor and platforms should be indicated. In the figure, Stair A is located between axes ①-② and Ⓑ-①/B. The horizontal dimension of the staircase is 3300mm×7800mm. There are two runs of the stair with 11 steps for each run. The width of the step is 300mm. The horizontal projection length of the bench is 300mm×11 = 3300mm. The width of the bench is 1675 (1475) mm. The size of the resting platform is 3300mm×2600 (1900) mm. The elevations of the ground floor, middle floors and the platforms are given. The elevation of the ground floor is ± 0.000. The elevation of the resting platform between 1st and 2nd floors is 2.100, whereas the elevation of that between 2nd and 3rd floors is 6.000.

The sectioning positions of the stair section shall be indicated in the ground floor stair plan. The lines used in the stair plan are the same as those in the corresponding architectural plan.

2) Stair section drawing

Figure 3.8.3 is the section drawing of Stair A. It is drawn according to the sectioning position and projection direction in the first floor plan of the stair. An imaginary vertical sectioning plan is used to firstly cut the stair into two parts and then the stair section is obtained by the projection method. In Figure 3.8.3, the stair structures on the 2nd floor and 3rd floor are totally the same, so it is only necessary to draw the sections of the ground floor, middle standard layer and top layer with broken lines. The elevations of the middle layers shall be added in brackets.

In the stair section drawing, the first run is cut with an imaginary plan and shall be filled. The second run may not be cut, so it is visible and contour lines of the second run is drawn.

The stair section drawing is used to display the elevations of each floor, the number of steps, methods of components, types and heights of handrails, elevations and sizes of door and window openings, etc. Therefore, the names, numbers and sizes of axes, elevations and sizes of floor, ground, platforms, and door and window openings shall be included. The height of stairs shall be denoted in the form of the number of steps in vertical direction × step height. The indexing symbols of handrails, breast boards and steps shall also be drawn.

The lines used in a stair section drawing are the same with those in the corresponding architectural section drawing.

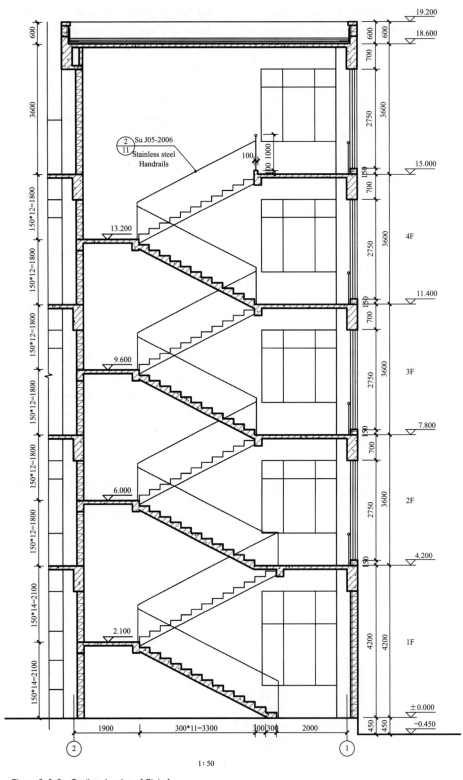

Figure 3.8.3 Section drawing of Stair A

Chapter 4
Structural Construction Drawings

A building comprises structural members (such as beams, columns, slabs, walls, and foundations) and building accessories (such as doors, windows, and balconies). Some main bearing members support each other to form a building structure. According to structure types, building structures may be classified into masonry structure, frame structure, shear wall structure, frame-shear wall structure, tube structure, etc. According to the types of structural materials, building structures may be classified into timber structure, concrete structure, reinforced concrete structure, steel structure, composite structure made of at least two materials, etc. The construction drawings of building structures with different materials have different graphic characteristics.

In the design of building structures, the first thing carried out by architects is to determine the arrangement and layout of the building according to the client's requirements. Structural engineers then determine the best structural system or form to bring the architect's concepts into being. Architects and engineers should work together in the design stage. Once the building form and structural arrangement have been finalized, the following items should be determined in the design stage: the idealization of the structure into load-bearing frames and members for analysis and design, estimation of loads, analysis and determination of maximum moments, thrusts, and shears for design, design of sections and reinforcement arrangements for slabs, beams, columns and walls, and preparation of structural construction drawings.

Traditional structural construction drawings generally include two main parts: structural layout and detail drawing of members. The structural layout shows the overall arrangement of members in the structure and the detailed drawing of members shows the shape, dimensions, and materials of each member.

A structural construction drawing is the main technical base for foundation trench excavation, formwork design, steel bar bending, pouring concrete, installation of a beam, column and slab, and construction organization.

4.1 Overview

1. Codes of common members

Structural members are manifold and complex. To simplify drawing contents, codes are used to represent the names of members such as beams, columns, and slabs. The code composed of the initials (first letter) of Pinyin of the Chinese names of members is followed by the model number or a serial number of the member. According to the ***Standard of Structural Drawings*** (GB/T 50105—2010), the code names of precast reinforced concrete members, the cast-in-situ concrete members, steel members, and timber members in Table 4.1.1 can be directly used.

In the drawings, except concrete members, the members of other materials should be labeled with material codes before the member code. Material codes should be explained in the drawing. For example, "Y-" should be added before the code of the prestressed concrete member. Y-DL indicates a prestressed reinforced concrete crane beam.

2. Introduction to reinforced concrete structures

Concrete is a mixture of sand, gravel, crushed rock, or other aggregates. Sometimes one or more admixtures are added to change the characteristics of concrete, such as workability, durability, and hardening time.

Like most rocklike substances, concrete has high compressive strength and low tensile strength. Reinforced concrete is a combination of concrete and steel bars that provides tensile strength. Steel bar is also capable of resisting compression forces and is used in columns as well as other structures.

1) Advantages of reinforced concrete as a structural material

Codes of common members   Table 4.1.1

No.	Names	Codes	No.	Names	Codes	No.	Names	Codes
1	Slab	B	19	Ring beam	QL	37	Cushion cap	CT
2	Roof slab	WB	20	Lintel	GL	38	Equipment foundation	SJ
3	Hollow slab	KB	21	Coupling beam	LL	39	Pile	ZH
4	Channel slab	CB	22	Foundation beam	JL	40	Retaining wall	DQ
5	Folded slab	ZB	23	Stair beam	TL	41	Trench	DG
6	Ribbed slab	MB	24	Frame beam	KL	42	Column bracing	ZC
7	Stair slab	TB	25	Frame-supported beam	KZL	43	Vertical bracing	CC
8	Cover slab or trench cover	GB	26	Roof frame beam	WKL	44	Horizontal bracing	SC
9	Baffle or eaves fascia	YB	27	Purlin	LT	45	Stair	T
10	Crane safety walkway slab	DB	28	Roof truss	WJ	46	Canopy	YP
11	Wallboard	QB	29	Bracket	TJ	47	Balcony	YT
12	Valley tray	TGB	30	Skylight truss	CJ	48	Beampad	LD
13	Beam	L	31	Frame	KJ	49	Built-in fitting	M-
14	Roof beam	WL	32	Rigid frame	GJ	50	End wall of the skylight	TD
15	Crane beam	DL	33	Support	ZJ	51	Reinforcing mesh	W
16	Monorail crane-beam	DDL	34	Column	Z	52	Reinforcement cage	G
17	Tracks connection	DGL	35	Frame column	KZ	53	Foundation	J
18	Car bumper	CD	36	Construction column	GZ	54	Embedded column	AZ

Reinforced concrete may be the most important construction material. It is used in almost all structures, such as buildings, bridges, pavements, dams, retaining walls, tunnels, drainage and irrigation facilities, tanks.

The universal construction material has the following advantages:

(1) It has a higher compressive strength per unit cost compared to most other materials.

(2) Reinforced concrete has excellent resistance to the actions of fire and water and is the best construction material in the water environment. During fires with average intensity, the members with a satisfactory cover of concrete over the reinforcing bars suffer only surface damages without any structural failure.

(3) Reinforced concrete structures are rigid.

(4) It is a low-maintenance material.

(5) Compared with other materials, it has a long service life. Under proper conditions, reinforced concrete structures can be used in-

definitely without the reduction of their load-carrying abilities. The strength of concrete does not decrease with time. On the contrary, it increases over a long period because of the lengthy solidification process of the cement paste.

(6) It is the only available economical material for foundations, floor slabs, basement walls, piers, and similar structures.

(7) Concrete can be cast into an extraordinary variety of shapes from simple slabs, beams, and columns to high arches and shells.

(8) In most applications, concrete takes advantage of inexpensive local materials (sand, gravel, and water) and requires a relatively small quantity of cement and steel bars, which may have to be shipped conveniently.

(9) A lower grade of skilled labor is required for the construction process compared with other materials such as structural steel.

2) Disadvantages of reinforced concrete as a structural material

To use concrete properly, the designer should be familiar with its disadvantages as follows:

(1) Concrete has insufficient tensile strength and requires the use of tensile reinforcement.

(2) Forms are required to hold the concrete in place until it hardens sufficiently. Besides, falsework or shoring may be necessary for keeping the molds in place for roofs, walls, floors, and similar structures until concrete members gain sufficient strength to support themselves. Formwork is expensive and formwork costs account for one-third to two-thirds (average of about 50%) of the total cost of a reinforced concrete structure. To improve the economy of reinforced concrete structures, the reduction of formwork costs is the key point.

(3) Due to the low strength per unit of weight of concrete, concrete members are generally heavy. This should be completely considered in long-span structures, where concrete's large deadweight has a significant effect on bending moments. Lightweight aggregates can be used to reduce concrete weight, but the cost of the concrete is increased by the use of lightweight aggregates.

(4) Similarly, due to the low strength per unit of volume of concrete, concrete members are relatively large. The large volume should be considered in tall buildings and long-span structures.

(5) The properties of concrete vary significantly with the proportion ratio and mixing parameters. Furthermore, the mixing or curing process of concrete is not performed as carefully as the production of other materials, such as structural steel and laminated wood.

Two characteristics of concrete (shrinkage and creep) may cause serious safety problems.

3) Compatibility between concrete and steel

Concrete and steel bars work together in reinforced concrete structures. The advantages of each material can compensate for the disadvantages of the other. For instance, the significant disadvantage of concrete is low tensile strength, but the high tensile strength is one of the significant advantages of steel. The tensile strength of reinforcing bars is approximately 100 times that of common concrete.

The two materials bond together well, thus decreasing the chance of slippage between them. They act together as a unit in resisting tensile forces. The obtained excellent bonding force is the consequence of the chemical adhesion between the two materials, the natural roughness of bars, and the closely spaced rib-shaped deformations on bar surfaces.

Reinforcing bars are subjected to corrosion, but surrounding concrete provides them with excellent protection. The strength of exposed steel subjected to the high temperature in fires of ordinary intensity is nil, but the reinforcing bars enclosed in concrete have satisfactory fire ratings. Finally, concrete and steel bars work well together during temperature changes because they have similar coefficients of thermal expansion. The coefficient of steel is 0.000012 per unit length per degree Centigrade, while it varies for concrete from about 0.00001.

4) Strength grades of concrete

Compressive strength is the most important property of concrete. The characteristic strength of the concrete grade is measured by the 28-day cube strength. Standard cubes of 150 or 100mm for aggregate not exceeding 25mm in size are crushed to determine the strength. According to **Code for Design of Concrete Structures** (GB 50010—2010), concrete strength is classified into 14 grades: C15, C20, C25, C30, C35, C40, C45, C50, C55, C60, C65, C70, C75 and C80. C50 ~ C80 belongs to high-strength concrete.

5) Types and grades of steel bars

According to chemical composition, steels used in concrete structures may be classified into carbon steel and ordinary low-alloy steel. According to the surface, steel bars can be classified into the plain bar and ribbed bar (the surface has herringbone or spiral patterns). Ordinary steel bars used in concrete structures may be hot-rolled bars. Prestressed steel bars used in concrete structures may be stress-relief steel wires, spiral ribbed steel wires, indented steel wires, stranded steel wires, or heat-treated steel bars.

Hot-rolled steel bars are classified into four grades according to their mechanic index. Their symbols and characteristic values of strength are shown in Table 4.1.2.

Characteristic values of ordinary steel bars   Table 4.1.2

Steel grade of bars	Symbols	Nominal diameter (mm)	Characteristic values of yield strength f_{yk}	Characteristic values of ultimate strength f_{stk}
HPB300	⏀	6 ~ 22	300	420
HRB400 HRBF400 RRB400	⏀ ⏀^{F} ⏀^{R}	6 ~ 50	400	540
HRB500 HRBF500	⏀ ⏀^{F}	6 ~ 50	500	630

6) Concrete cover and steel hooks

A certain distance from the outer surface of the outermost bar to the concrete member should be maintained to protect steel bars from rust, fire, and corrosion. In other words, the concrete cover with a certain thickness should be considered in the construction.

The concrete covers for ordinary steel bars and prestressing bars should meet the following requirements:

(1) The distance from the concrete cover to main bars in members should be no less than the nominal diameter of bar d.

(2) The minimum concrete cover in Table 4.1.3 applies to concrete structures whose design working life is 50 years. As for the structures whose design working life is 100 years, in the Type I environment, the distance from concrete cover to the outermost bar should be no less than 1.4 times the value in Table 4.1.3. In Types II and III environments, specific measures shall be considered in the design.

(3) When the concrete grade is no greater than C25, the thickness of the concrete cover shown in Table 4.1.3 should be increased by 5mm.

(4) As for the bottom reinforcement of foundations, when a concrete cushion exists, the concrete cover should be measured from the top of the cushion and should be no less than 40mm.

Minimum concrete cover (mm)
Table 4.1.3

Environment types	Slabs and walls	Beams and columns
I	15	20
II$_a$	20	25

Continued

Environm- ent types	Slabs and walls	Beams and columns
II$_b$	25	35
III$_a$	30	40
III$_b$	40	50

To guarantee the good bonding between steel bar and concrete, a hook of 180° should be used at the end of plain bars. However, a hook of 180° should not be used at the end of ribbed bars because of the strong bonding between ribbed bar and concrete. The stirrup should be made into a hook of 135° at the junction.

7) Names and functions of steel bars

According to different functions of steel bars in members, steel bars may be classified into the concrete slab, beam, and column, as shown in Figure 4.1.1.

Longitudinal bars are the main steel bars used in concrete members to bear tension or pressure. They are arranged in concrete beams, columns, slabs, and other members.

Stirrups are steel bars bearing shearing force or torque in concrete members and are used to fix the locations of longitudinal bars in concrete beams and columns.

Erection bars form the reinforcement framework together with longitudinal bars and stirrups and are generally used to fix the locations of steel bars in the beam.

Distribution bars are generally used in the slabs, and vertically arranged together with longitudinal bars in slabs to form the bar mesh. In addition, the load is uniformly transferred to longitudinal bars through distribution bars. They can resist the temperature deformation caused by thermal expansion and contraction.

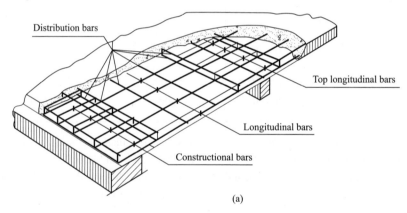

(a)

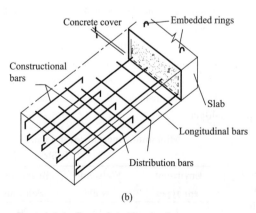

(b)

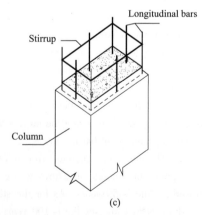

(c)

Figure 4.1.1 Types of steel bars(one)
(a) Continuous slab;(b) Simple support slabs;(c) Column

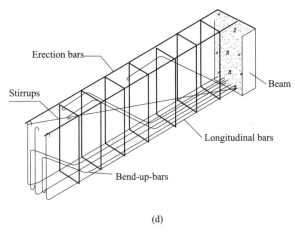

Figure 4.1.1 Types of steel bars(two)
(d) Beam

Side bars should be placed on both sides along with the depth of beam when h_w is no less than 450mm.

Torsion bars should be placed on both sides along with the depth of the beam when the beam works under a torsional moment. Side bars can be replaced by torsion bars when the diameter of torsion bars is longer than that of side bars.

Bend-up bars are designed for increasing the shear resistance.

Hangers are a kind of steel bars that transfer the concentrated force acting on the bottom of the concrete beam to the top and improve the shear resistance of the beam under concentrated load.

Constructional bars are a kind of steel bars that take into account all kinds of immeasurable factors in reinforced concrete members. Moreover, they are placed according to the mandatory requirements of related structure design standards.

8) Annotation methods of steel bars

Two annotation methods of steel bars are provided as follows:

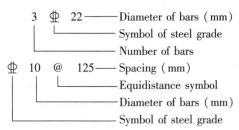

9) Anchorage and splice of steel bars

Steel bars must be reliably anchored in concrete. The anchorage length of longitudinal bars in concrete is affected by concrete strength grade and the diameter, type, and shape of bars.

The splices of bars can be classified into lapping, mechanical, and welding splices.

10) Graphical presentation method of reinforced concrete members

To clearly show reinforcement bars in members, concrete is assumed to be transparent so that reinforcement bars in members are visible in construction drawings. A reinforcement drawing is mainly used to show the arrangement of steel bars. In a reinforcement drawing, a bold solid line is used to denote reinforcement bars along the length direction, and a small solid circle is used to denote the cross-section of reinforcement bars. The profile of the members is drawn with a middle solid line or a thin solid line.

3. Ichnographic representation method of structural construction drawings

In the ichnographic representation method of structural construction drawings (IRM), the design representation method of concrete structural construction drawings in China has been significantly modified. It has been listed in the ***Key Promotion Projects of National " Ninth Five-Year" Science and Technology Achievements***

(Project No.: 97070209A) by the State Scientific and Technological Commission and the **Key Promotion Projects of Science and Technology Achievements** in 1996 (Project No.: 96008) by Ministry of Housing and Urban-rural Development.

Ichnographic representation, in general, is to show dimensions and reinforcement details of structural members directly in the structural plan layout according to the ichnographic representation method. A completed set of structural construction drawings shall include both IRM drawings and standard detail drawings.

Generally speaking, the ichnographic representation method is to directly express the size and reinforcement bars of structural members in the structural layout plans of various members according to the IRM. Then, a set of new and complete structural designs is obtained after the layout plans are integrated with standard structural drawings. This method is different from the traditional representation method, in which members are indexed from the structural layout plans and then reinforcement detail drawings are plotted one by one.

When structural construction drawings are plotted with IRM according to the design requirements of a specific project, the dimensions and reinforcement details of structural members in the plan layout drawings should meet the relevant requirements of IRM. When construction drawings are prepared, the full set of construction drawings should be arranged in the following order: foundations, columns, shear walls, beams, slabs, stairs, and other members.

Dimensions and reinforcement details of structural members can be noted in the plan layout according to the following three methods: the ichnographic annotation method, tabulated annotation method, and sectional annotation method. This chapter mainly introduces the requirements of ichnographic construction drawings of reinforced concrete foundations, columns, beams and slabs, and slab-stairs.

4. Comparison of reinforced concrete and structural steel for buildings

When a particular type of structure is being considered in the design, a student may have a question, "Should reinforced concrete or structural steel be used?" There is much joking on this point, with the proponents of reinforced concrete referring to steel as that material that rusts and those favoring structural steel referring to concrete as the material that, when overstressed, tends to return to its natural state—that is, sand and gravel.

The question has not a definite answer since both materials have corresponding advantages when they are utilized in many types of structures. They are often used in the same structures with satisfactory results.

The selection of structural materials for a building depends on the height and span of the structure, the material market, foundation conditions, local building codes, and architectural considerations. As for buildings of less than 4 stories, reinforced concrete, structural steel, and wall-bearing construction are competitive. As for buildings of 4 to 20 stories, reinforced concrete and structural steel are economically competitive materials. Structural steel has been used in many buildings of more than 20 stories. Today, however, reinforced concrete is increasingly applied in buildings of more than 20 stories and several higher reinforced concrete buildings have been completed in the world.

Foundation conditions often affect the selection of materials for the structural frame. If foundation conditions are poor, a lighter structural steel frame is recommended. The building code in a particular city may favor one material. For instance, many cities have fire zones in which only fireproof structures can be erected, so reinforced concrete is recommended. In addition, the construction period of structural steel frames is short as they can be erected more quickly than reinforced concrete ones. The time advantage, however, is not so important in some cases. For example, if the structure has a fire-resistance rating, the steel frame should be covered with some fire proofing materials after it is erected.

4.2 General Requirements in Standard for Structural Drawings of Buildings

This section mainly introduces general requirements in *Standard for Structural Drawings* (GB/T 50105—2010), including the general requirements of lines, scales, and structural drawings of reinforced concrete.

1. Lines in structural drawings of buildings

According to the complexity and scale of structural drawings, the basic line weight b is firstly selected from Table 4.2.1, followed by the weight group. In a drawing, the same line weight group should be adopted for different components with the same scale.

2. Scale

Scales may be selected from common scales in Table 4.2.2 for structural construction drawings according to the purposes of drawings and the structural complexity. The optional scales can also be selected in special cases. When the longitudinal and transversal section sizes of members vary greatly, different scales can be selected in the same detail drawing. Axis dimensions and member dimensions vary with scale.

3. General representation method of steel bars

The general representation method of steel bars shall meet the requirements in Table 4.2.3.

4. Drawing methods of steel bars (Table 4.2.4)

5. Representation methods of steel bars in a plan, elevation, and section

The arrangement of steel bars in a plan shall be represented with the method shown in Figure 4.2.1. When the space is not enough for

Lines   Table 4.2.1

Names		Line Type	Line weight	General uses
Solid line	Bold	———	b	Bolt; main steel bar line; single-line structural member line, steel or wooden support and tie bar line, and horizontal line and sectional line under the drawing name in a structural plan
	Semi-bold	———	$0.7b$	Sectioned or visible wall profile, foundation profile, steel structure profile, timber structure profile, and steel bar line in a structural plan and a detailed drawing
	Middle	———	$0.5b$	Sectioned or visible wall profile, visible profile of reinforced concrete members, and steel bar line in a structural plan and a detailed drawing
	Thin	———	$0.25b$	Leader line for dimensions, elevation symbol line, index symbol line, and dimension line

Continued

Names		Line Type	Line weight	General uses
Dash line	Bold	-------	b	Invisible steel bar and bolt line; invisible single-line structural member line and steel or wooden support line in a structural plan
	Semi-bold	-------	$0.7b$	Invisible profile of the member, wall, steel structure, timber structure, and invisible steel bar line in a structural plan
	Middle	-------	$0.5b$	Invisible profile of the member, wall, steel structure, timber structure, and invisible steel bar line in a structural plan
	Thin	-------	$0.25b$	Pipe trench profile and invisible profile of reinforced concrete members in a foundation plan
Long-dash-dot line	Bold	—·—	b	Column bracing, vertical bracing, and the centerline of the axis drawing of the equipment base
	Thin	—·—	$0.25b$	Positioning axis, symmetric line, centerline, the centerline of gravity
Long-dash double-dot line	Bold	—··—	b	Prestressed steel bar line
	Thin	—··—	$0.25b$	Outline of the original structure
Break line		—⋀—	$0.25b$	Boundary line
Wavy line		～～	$0.25b$	Boundary line

Scales   Table 4.2.2

Drawing names	Common scales	Optional scales
Structure plan and foundation plan	1 : 50, 1 : 100, 1 : 150	1 : 60, 1 : 200
Ring beam plan, pipe trench, and underground facilities in general layout drawing	1 : 200, 1 : 500	1 : 300
Detail drawing	1 : 10, 1 : 20, 1 : 50	1 : 5, 1 : 25, 1 : 30

marking steel bars, leader lines can be used. The leader lines shall be drawn with a middle solid line or a thin solid line. When the layout of members is relatively simple, the structural layout plan and slab reinforcement plan can be drawn jointly in a drawing.

General representation method of steel bars   Table 4.2.3

No.	Names	Legends	Descriptions
	Ordinary steel bars		
1	Cross-section of steel bars	•	—
2	Steel bar end without hook	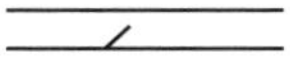	The lower figure shows that if the projections of long and short steel bars are overlapped, the end of the short steel bar is indicated with a slash with an angle of 45°
3	Steel bar end with a hook of 180°		—
4	Steel bar end with a hook of 90°		—
5	Steel bar end with screw thread		—
6	Steel bar lapping without hook		—
7	Steel bar lapping with a hook of 180°		—
8	Steel bar lapping with a hook of 90°		—
	Prestressed steel bars		
9	Prestressed steel bars or stranded steel wire		
10	Section of post-tensioning prestressed steel bars and unbonded prestressed steel bars	$\oplus$	
11	Section of prestressed steel bars	$+$	
	Reinforcing mesh		
12	Plan of a reinforcing mesh	 W-1	
13	Plan of a row of reinforcing meshes with the same size	 3W-1	

Drawing methods of steel bars   Table 4.2.4

No.	Descriptions	Legends
1	If double-layer steel bars are configured in a floor slab, the hooks of the bottom steel bars shall be bent upward or toward to the left, but the hooks of the top steel bars shall be bent downward or toward the right	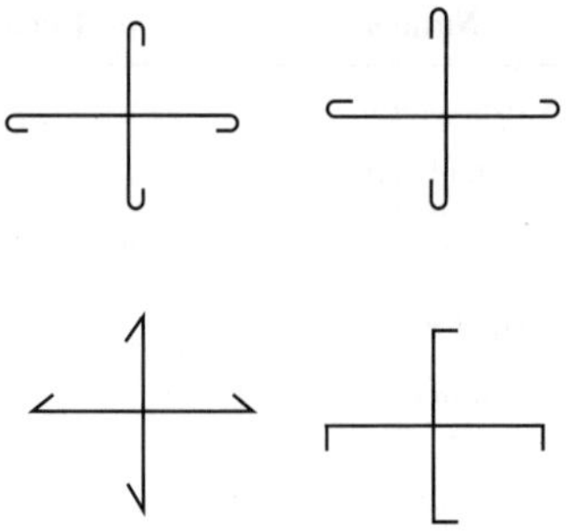 (Bottom layer)　(top layer)
2	If double-layer steel bars are configured in a reinforced concrete wall, the hooks of YM steel bars shall be bent in the upward or left direction, whereas the hooks of JM steel bars shall be bent in the downward or right direction (JM: near side; YM: far side)	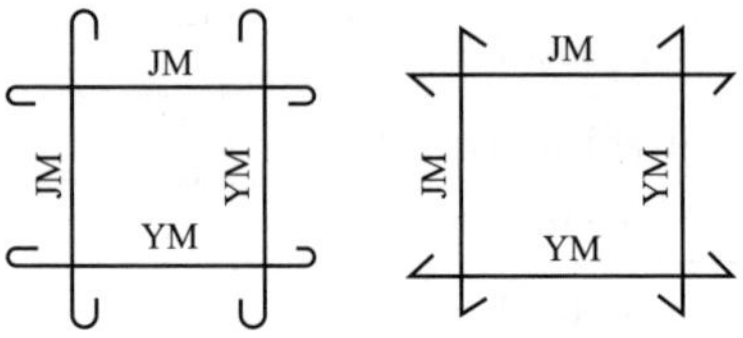
3	If the steel bars layout can not be expressed clearly in the sectional drawing, a detailed drawing shall be added (reinforced concrete walls, stairs, etc.)	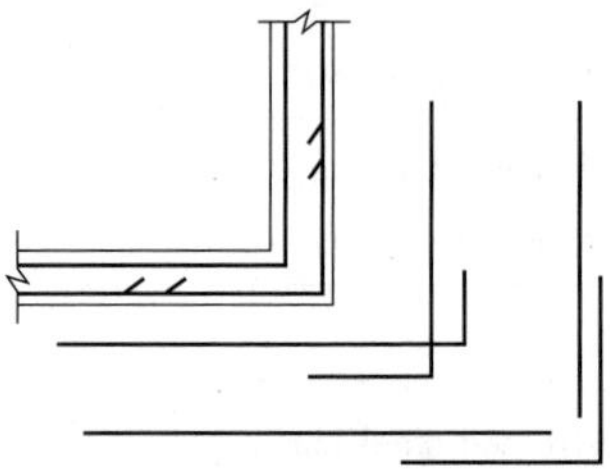
4	If the layout of stirrups is complex, the steel bar detail drawing and descriptions shall be added	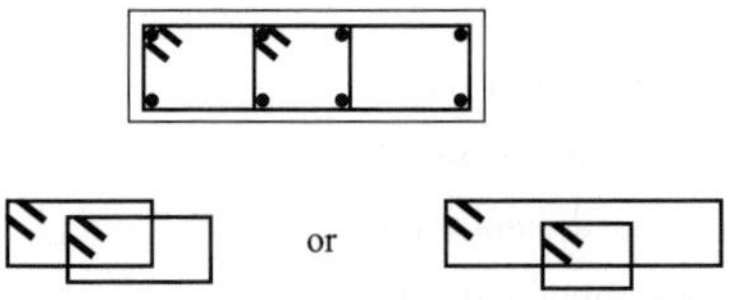 or
5	In each group, the same steel bars and stirrups can be denoted with a bold solid line and the remaining steel bars can be denoted with a horizontal thin line with two short bold solid lines at their two ends	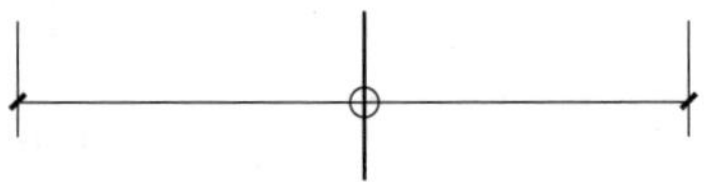

The arrangement of steel bars in elevation and cross-section shall be represented with the methods shown in Figure 4.2.2.

6. Simplified representation method of steel bars

As for symmetrical reinforced concrete members, in the same drawing (Figure 4.2.3), half of the member is used to represent the

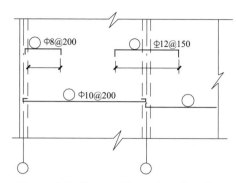

Figure 4.2.1　Representation method of steel bars in the reinforcement plan of slabs

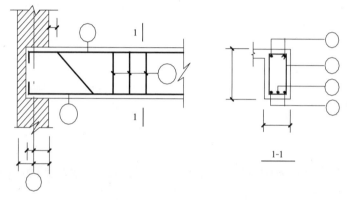

Figure 4.2.2　Reinforcement drawing of a beam

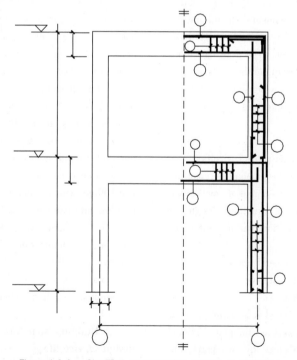

Figure 4.2.3　Simplified representation method of reinforcing drawing of symmetrical members

formwork, whereas the other half is used to represent reinforcement bars.

When reinforcement bars of reinforced concrete members are relatively simple, the reinforcement plan is drawn according to the following requirements. At the lower-left corner of the formwork of the spread footing, the wavy line is drawn to indicate the place where the reinforcement bars are drawn and the diameter and spacing of reinforcement bars are marked (Figure 4.2.4a). A wavy line is drawn at a particular place of other members, where the reinforcement is drawn, and the diameter and spacing of reinforcement bars are marked (Figure 4.2.4b).

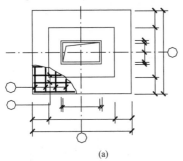 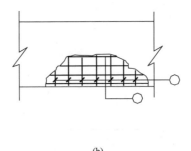

(a) (b)

Figure 4.2.4 Simplified representation method of reinforcing drawing of reinforced concrete members
(a) Spread footing;(b) Other members

4.3 General Notes of Structure Design

In the construction drawing design stage, the structure design documents should include the drawing catalog, design notes, design drawing, and calculation book. The drawing catalog shall be arranged according to the serial number of drawings. The new drawing shall be included first, followed by an optional drawing and standard drawing.

General notes of structural design shall be compiled for every single project. Unified general notes of structural design shall be compiled for multiple sub-projects. If it is a small single project, general notes of structural design can be written on the foundation plan and the structural plan of each floor.

Examples of the general notes of structural design are provided as follows:

1. Project overview:

This project is located at a sloping field in Addis Ababa and belongs to Ethiopian Airlines. The building is constituted by two ports: the main hotel building and the banqueting hall.

The main building is 1567m long and 96.1m wide and composed of one underground floor and 8 above-ground floors. The underground floor is 4.7m high and upper ground floor (UG floor) each floor 7.2m, 1-7 floor, each floor is 3.6m, total height 37.1m. The port of the main building is the frame shear wall structure. The transfer floor is on the above-ground floor. The banquet hall is 66m long and 58m wide. Only the underground floor and above-ground floor are reinforced concrete frame structures. There is a 100mm wide seismic joint between the main building and banquet hall.

The foundation of the whole building is the flat plate raft foundation. The thickness is respectively 1000mm and 600mm. During the formwork supporting, some column stubs at the bottom of some columns should be carefully considered.

2. Building structure safety grade and design service life:

1) Building structure safety grade: Grade II.

2) Design service life: 50 years.

3) Ground foundation design grade: Grade Ⅱ.

4) Main hotel building. The seismic grade of the support frame is Grade Ⅱ. The seismic grade of the reinforced part of the shear wall is Grade Ⅲ. The seismic grade of the normal part of the shear wall is Grade Ⅳ. The shear wall below elevation 15.8m is a strengthening part of the shear wall. The seismic grade of the reinforced concrete frame of the banquet hall is Grade Ⅲ.

3. **Natural conditions:**

1) Fundamental wind pressure: $0.35kN/m^2$; ground roughness: Grade B.

2) Site earthquake fundamental intensity: 6 degrees ($0.05g$).

3) Building site grade: Type I.

4) Geological conditions of the construction site.

4. **Drawing instruction.**

5. **This project is designed according to related standards, codes, and rules.**

6. **Design calculation program.**

7. **The standard value of the evenly distributed live load is designed.**

8. **Ground foundation.**

9. **Main structure material.**

10. **Columns, walls and beams.**

11. **Structure plan.**

12. **Infilled wall.**

13. **Others.**

4.4 Ichnographic Construction Drawings of Foundations

The foundation is the underground load-bearing structure under the load-bearing wall or column. It bears the full load of the building and transfers it to the subsoil.

According to the bearing of the upper part and the bearing capacity of the subsoil, building foundations may be classified into spread footing under the column, strip foundations under the wall (or column), cross strip foundation under the column, raft foundations, and box foundations. According to foundation materials, building foundations may be classified into masonry foundations, concrete foundations, and reinforced concrete foundations.

Figure 4.4.1 shows common foundation types. Figure 4.4.1(a) shows a strip foundation under a load-bearing wall. A Foundation pit is a soil pit excavated on the ground for foundation construction. The subsoil is a natural or reinforced soil layer beneath the foundation. The foundation wall is the buried underground part of the bearing wall. Masonry between the foundation wall and the cushion is called the pedestal foot and a layer under the pedestal foot is called the cushion. The buried depth of the foundation refers to the distance from the outdoor ground to the bottom of the foundation. Figure 4.4.1(b) shows spread footing with stepped section. Spread footing is connected with the foundation beam and the load-bearing walls or partition walls are built on foundation beams.

A foundation drawing is to display the plan layout and details of the foundation under the ground. It is the basis for a construction setting out, foundation pit excavation, and construction. Traditional foundation drawings usually include foundation plan layout and foundation detail drawing.

1. Traditional representation methods

Traditional representation of foundations includes two parts: foundation plan layout and foundation detail drawing.

The foundation plan layout denotes the plan drawing when the foundation pit is not filled with earth. In other words, it is a horizontal section view that cuts through the walls of the foundation.

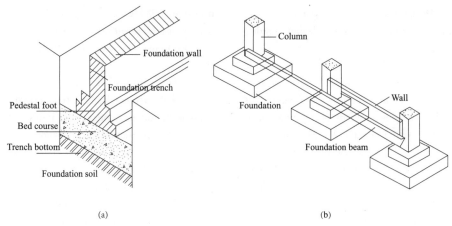

Figure 4.4.1 A typical foundation
(a) Strip foundation; (b) Spread footing

The foundation plan layout denotes the layout of the foundation when the foundation pit is not backfilled. Under the first floor, the building is cut with an imaginary horizontal cutting plane to remove the part above the cutting plane and orthographically project the remaining part to the horizontal plane. Finally, the foundation plan layout is obtained.

1) Strip foundation plan

Figure 4.4.2 shows the foundation plan layout of an experimental building of a middle school. This building is a multi-storey brick-concrete structure made of a brick wall, reinforced concrete floor, and roof. The foundation under the brick wall is a reinforced concrete strip foundation.

(1) Drawing name, scale, and positioning axis

The drawing name is the foundation plan layout. The scale usually is the same as that of the floor plan, 1 : 100. The positioning axis is to determine the positions of the foundation wall and columns. The serial numbers must be consistent with those in the floor plan. The number in the horizontal direction is 1 to 5 and the number in the vertical direction is A to D.

(2) Foundation plan layout

A foundation plan layout contains foundation wall, construction column, frame column, foundation beam, foundation ring beam, and profile of foundation.

In the foundation plan layout, on both sides of the axis, the profile of the foundation wall is drawn with a bold line. The material legends are the same as those in the floor plan. The legend of the brick wall may be omitted in the foundation plan layout with a scale of 1 : 100. Generally, the reinforced concrete construction column (GZ1-GZ3) and the column of the canopy (KZ1-KZ3) are drawn with bold solid lines. The sectional dimensions of the construction column are provided as follows: GZ1 (240mm×240mm), GZ2 (240mm×360mm), and GZ3 (360mm×360mm). The sectional dimension of frame columns is 400mm×400mm. The visible profile of the foundation is drawn with the middle solid line.

According to the seismic requirements, the foundation ring beam JQL shall be arranged in the foundation wall to prevent differential settlements. The foundation beam (JL1-JL7) is often arranged in the strip foundation at the location of a larger door opening to make the foundation bear the subgrade resistance. The projection of the foundation ring beam and foundation beam coincides with the projection of the wall, so the position of the centerline of beams is denoted with a bold long-dash dot line.

A foundation plan layout often contains the cross-sections of the foundation wall, construction columns, bearing columns, and the foundation profile. The detailed projection of the foun-

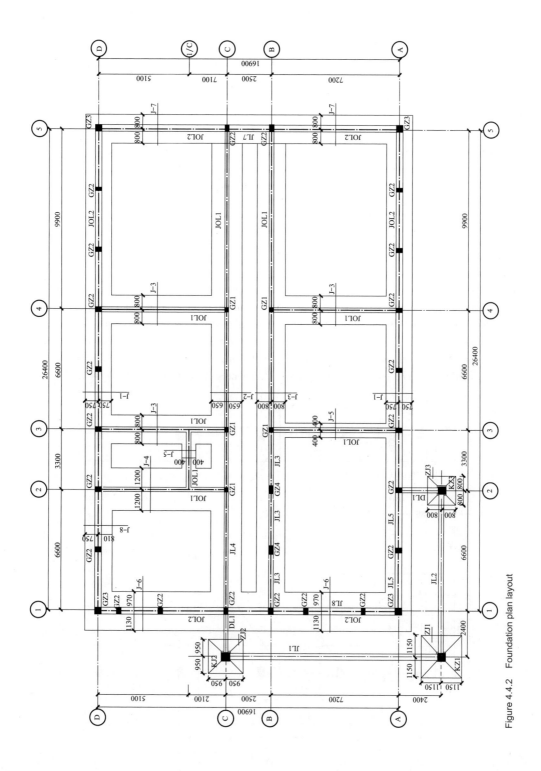

Figure 4.4.2 Foundation plan layout

dation (such as shapes, materials, and composition of foundation beams) is reflected in the foundation detail drawing.

(3) Dimensioning

The dimensioning contents in the foundation plan layout include positioning dimensions detail dimension, etc. Positioning dimensions mainly refer to the dimension between positioning axes, the dimension from axes to the foundation pit, and the dimension from axes to the foundation wall. The detailed dimensions of foundations mainly include the width of the foundation wall and the dimensions of the column section and foundation. The dimensions may be directly marked in the foundation plan layout or described with texts and codes.

2) Spread footing plan layout

Reinforced concrete frame structure refers to the bearing structure system composed of a frame beam, frame columns, and reinforced concrete slab. Reinforced concrete spread footing is commonly used under frame columns.

In the foundation plan layout in Figure 4.4.3, the cross-section of the column is shown and its material legend is denoted in black. The bottom shape of the spread footing and the top shape of the foundation beam are the visible profile, which is marked with solid middle lines. The positioning dimensions of columns, spread footings, and foundation beams, as well as the codes of the foundation (J1-J3) and the foundation beam (JL1, JL2), are also noted in the drawing.

3) Foundation detail drawing

A foundation plan layout only contains the foundation layout, but the shape, material, size, or details of each part of the foundation is not denoted. Therefore, a detailed drawing is required. A foundation detail drawing is used to display the cross-sectional shape, dimensions, and materials in detail. A foundation detail drawing is usually drawn according to different numbers of the foundation plan layout.

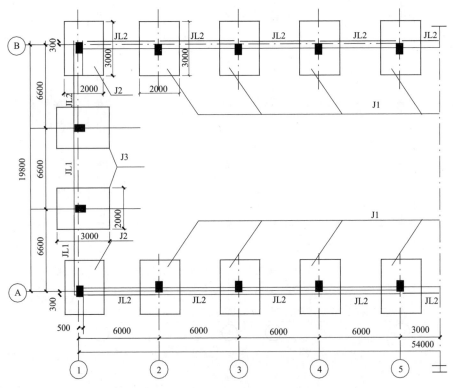

Figure 4.4.3 Spread footing plan layout

Figure 4.4.4 is the detailed drawing of spread footing under frame columns. Each spread footing has a consistent sectional shape and reinforcement form, so it is necessary to draw one cross-section of the foundation. In addition, a chart is drawn to provide auxiliary descriptions.

In the detailed drawing of spread footings, the vertical section and plan layout is generally adopted. As shown in Figure 4.4.4, the foundation shape is denoted in the plan layout and the sectional form of the foundation and reinforcement bars in foundation slabs are denoted in the vertical section drawing. There is a 100mm thick concrete cushion below. To denote the bidirectional reinforcement mesh in the foundation slabs in the plan layout, wavy lines can be used at one corner.

2. Ichnographic construction drawing of spread footings

Spread footing is a basic form of the foundation under a column. Such foundation is classified into normal spread footing and socket spread footing depending on the construction way of columns (cast/precast). The cross-sections of foundation slabs are classified into stepped and slope sections, as shown in Figure 4.4.5.

1) Representation method of ichnographic construction drawing of spread footing

Ichnographic construction drawing of spread footings includes ichnographic annotation and section annotation. Ichnographic annotation and/or section annotation methods can be used in the design of spread footing construction drawings according to the specific situation of a project.

(1) In the plan layout of spread footings, the plan of spread footing should be combined with columns supported by foundations. If there are connection beams between foundations, they may be shown in a foundation layout plan or a separate drawing.

(2) The size of the foundations should be noted in the foundation plan layout. If the centerline of the socket or column of the spread footing does not coincide with the gridline in architectural drawings, the distance should be noted. If the size and numbering of foundations are the same, it is only necessary to denote one of them.

2) Numbering of spread footings

The numbers of spread footings are shown in Table 4.4.1.

3) Ichnographic annotation method of spread footings

The ichnographic annotation method of spread footings includes general annotation and specific annotation.

(1) General annotation

General annotations of spread footings include three compulsory items to be denoted in the foundation plan: foundation numbers, vertical dimensions of cross-section, and reinforcement details. There are also two optional items: elevation of the base of foundations (if the elevation difference exists) and text annotations. General annotations of normal spread footings are shown in Table 4.4.2 and general annotations of socket spread footings are shown in Table 4.4.3.

(2) Specific annotation

Dimensions of spread footings are shown in the plan layout of the foundation. For different foundations with the same number, only one of them shall be denoted with the specific annotation. If the plan layout of foundations is too small, the denoted one can be enlarged according to a proper scale. On other foundations with the same foundation number, only the foundation number should be denoted. Specific annotations of normal spread footings are shown in Table 4.4.4 and specific annotations of socket spread footings are shown in Table 4.4.5.

(3) Ichnographic annotation method of multi-column spread footings

Spread footings generally refer to single-column spread footings, but it may also be multi-column spread footings (two-column or four-column) Annotation methods of numbering, dimensions, and reinforcement details of multi-column spread footings are the same as those of single-column spread footings.

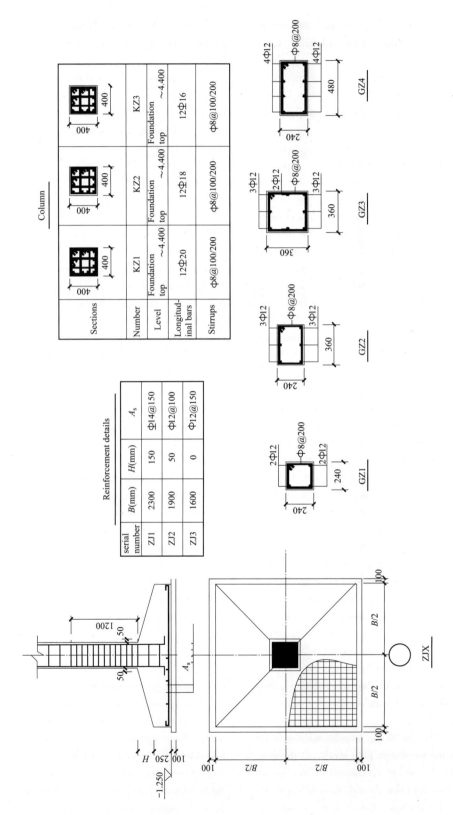

Figure 4.4.4　Detail drawing of reinforced concrete spread footing

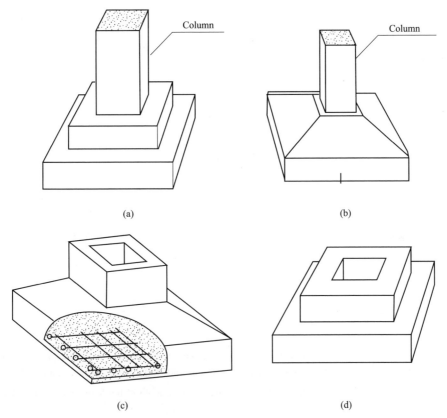

Figure 4.4.5 Common spread footing
(a) Normal spread footing with stepped section;(b) Normal spread footing with sloped section;
(c) Socket spread footing with sloped section;(d) Socket spread footing with stepped section

Numbers of spread footings   Table 4.4.1

Types	Cross-sections of foundation slabs	Type codes	Serial numbers
Normal spread footing	Step	DJ_J	××
	Slope	DJ_P	××
Socket spread footing	Step	BJ_J	××
	Slope	BJ_P	××

General annotations of normal spread footings   Table 4.4.2

Serial number	Denotation contents	Descriptions
1	Numbering of spread footing(compulsory) $DJ_J××$; $DJ_P××$	DJ_J01; DJ_P12 See Table 4.4.1

Continued

Serial number	Denotation contents	Descriptions
2	**Vertical dimensions of cross-section (compulsory)** $h_1/h_2/\cdots\cdots$	(1) A foundation with a stepped section. 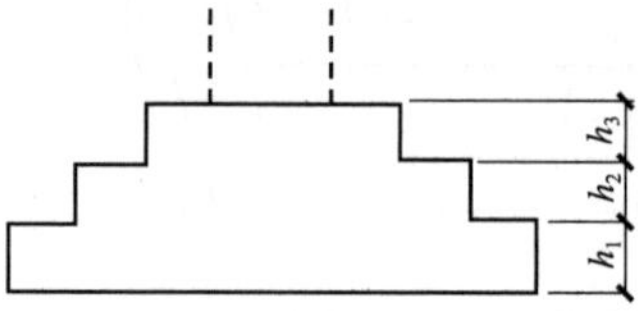 **e. g.** If vertical dimensions of normal spread footings with stepped cross-section are noted as 400/300/200, it is indicated that $h_1 = 400$, $h_2 = 300$, $h_3 = 200$. Therefore, the total depth of the foundation slab is 900. (2) A foundation with a sloped section. 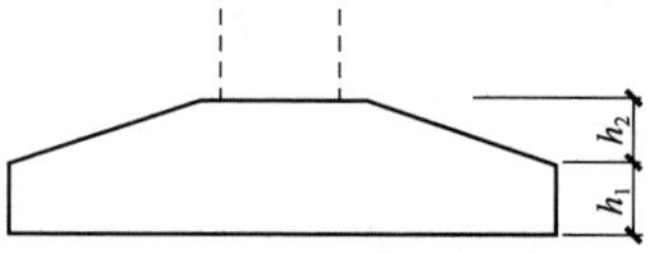 **e. g.** If vertical dimensions of normal spread footings with a sloped cross-section are noted as 400/200, it is indicated that $h_1 = 400$ and $h_2 = 200$. Therefore, the total depth of the foundation slab is 600
3	**Reinforcement details (compulsory)** (1) reinforcement details of foundation slabs B: X:⌽xx@xxx, Y:⌽xx@xxx	(1) "B" is used to represent the bottom reinforcement in slabs of all kinds of spread footings. (2) Prefix "X" is used to represent reinforcement in the direction X; Prefix "Y" is used to represent reinforcement in the direction Y. "X&Y" is used to represent the same reinforcement arranged in both directions. 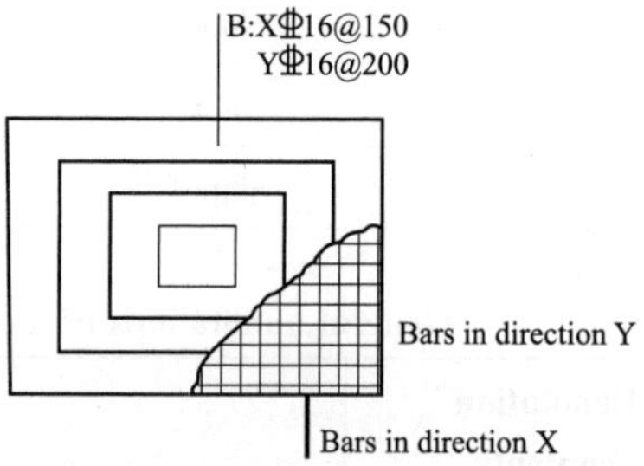 B:X⌽16@150 Y⌽16@200 Bars in direction Y Bars in direction X e. g. As shown in the above figure, the steel grade of reinforcement placed at the bottom of the foundation slab is HRB400. The diameter of bars in the direction X is 16 with the spacing of 150mm. The diameter of bars in the direction Y is 16 with a spacing of 200mm

Continued

Serial number	Denotation contents	Descriptions
3	(2) Vertical dimensions and reinforcement details of short columns of standard spread footing. (If the embedment depth of spread footing is large, reinforcement details of short columns should be denoted in spread footing). DZ xx⌽xx/xx⌽xx/xx⌽xx ⌽xx@xx -x.xxx~-x.xxx	(1) DZ is used to represent short columns of standard spread footing. (2) Longitudinal bars in short columns should be noted first, followed by stirrups and the elevation range of short columns. They should be noted as "corner longitudinal bars/middle longitudinal bars on the longer side/middle longitudinal bars on the shorter side, stirrups, elevation range of short columns". If the cross-section of a short column is a square, the reinforcement information can be denoted as "corner longitudinal bars/middle longitudinal bars in the direction X/ middle longitudinal bars in the direction Y, stirrups, elevation range of the short column". 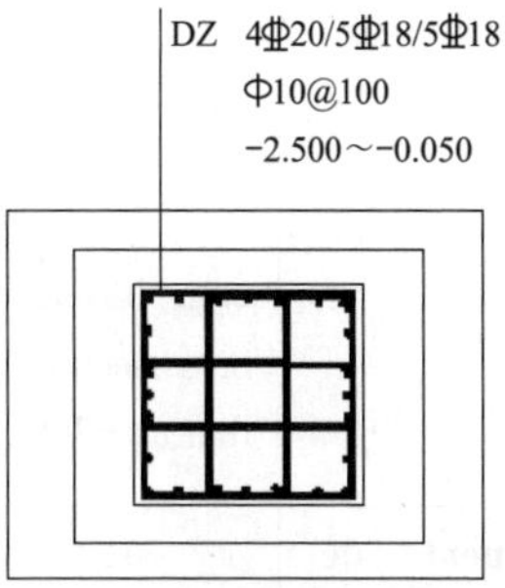 DZ 4⌽20/5⌽18/5⌽18 ⌽10@100 -2.500~-0.050 e.g. As shown in the above figure, the short column of spread footings is within the elevation range from -2.500 to -0.050. The steel grade of vertical longitudinal bars is HRB400 and the steel grade of stirrups is HPB300. Vertical longitudinal bars are corner bars (4 ⌽20) and middle bars (5 ⌽18) in direction X and middle bars (5 ⌽18) in direction Y. The diameter of the stirrups is 10 and the spacing is 100
4	Elevation of the base of the foundation (optional) (x.xxx)	If the elevation of the base of spread footings is different from the reference elevation of the base of the foundation, it should be denoted in brackets
5	Text annotations if necessary (optional)	If there are special requirements for spread footing in the design, it is recommended to provide an additional note

General annotations of socket spread footings Table 4.4.3

Serial number	Denotation contents	Descriptions
1	**Numbering of spread footing (compulsory)** $BJ_J\times\times; BJ_P\times\times$	$BJ_J01; BJ_P12$ See Table 4.4.1
2	**Vertical dimensions of cross-section (compulsory)** $a_0/a_1, h_1/h_2/\cdots\cdots$	(1) A foundation with a stepped section. 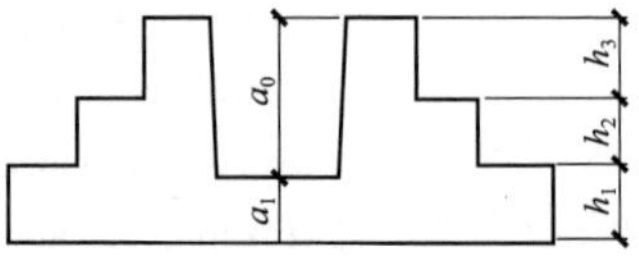 **e.g.** If vertical dimensions of socket spread footings with stepped cross-sections are denoted as 750/250, 300/300/400, it is indicated that $a_0=750, a_1=250, h_1=300, h_2=300$, and $h_3=400$. (2) A foundation with a sloped section. 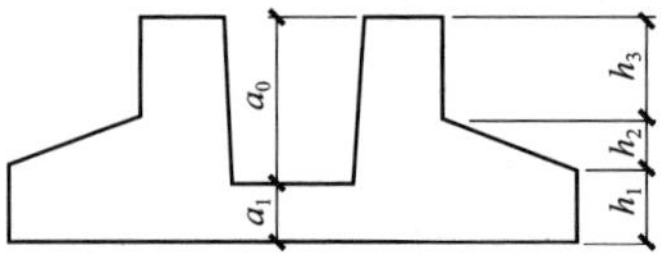 **e.g.** If vertical dimensions of socket spread footings with a sloped cross-section are denoted as 700/300, 400/200/400, it is indicated that $a_0=700, a_1=300, h_1=400, h_2=200$, and $h_3=400$
3	**Reinforcement details (compulsory)** (1) Reinforcement details of foundation slabs B: X: $\Phi\times\times@\times\times\times$, Y: $\Phi\times\times@\times\times\times$	See Table 4.4.2
	(2) Welded steel fabric at the top of the socket of spread footing. Sn $\times\times\Phi\times\times$	Sn is used to represent reinforcement bars on each side of the welded steel fabric at the top of the socket. 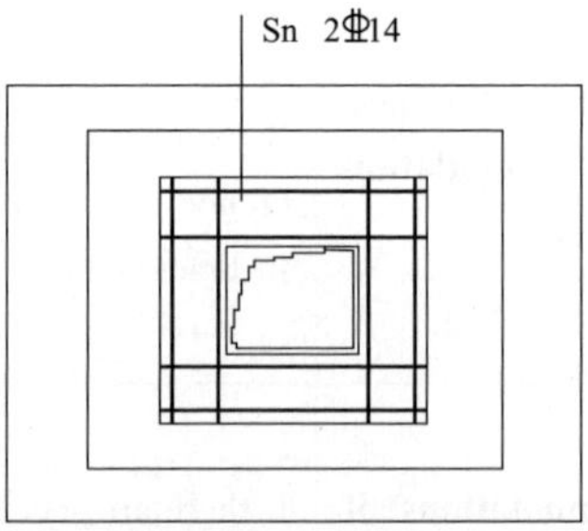 Sn 2Φ14 **e.g.** As shown in the above figure, the steel grade of welded steel fabric is HRB400. The diameter of the bars is 14 and two bars are arranged on each side

Continued

Serial number	Denotation contents	Descriptions
3	(3) Reinforcement bars in short columns of high-socket spread footings. (Also applicable to reinforcement bars in the socket walls of socket spread footings) O: ××ϕ××/ϕ××@ ××/ ϕ××@ ×× ϕ××@ ××/××	(1) O is used to represent reinforcement bars in short columns. (2) Longitudinal bars in short columns should be denoted first, followed by stirrups. They should be denoted as "corner bars/middle bars on the longer side/middle bars on the shorter side, stirrups (two spacing arrangements)". If the cross-section of a short column is a square, the reinforcement information may be denoted as "corner longitudinal bars/middle longitudinal bars in the direction X/ middle longitudinal bars in the direction Y, stirrups (two spacing arrangements: stirrups in the socket of short columns/stirrups in other parts)". 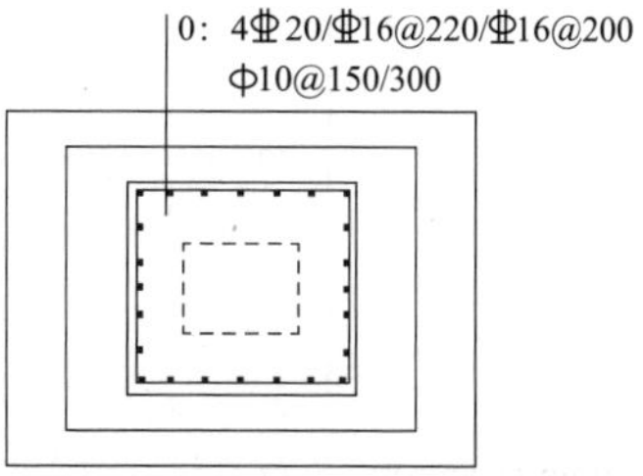 O: 4ϕ20/ϕ16@220/ϕ16@200 ϕ10@150/300 e. g. As shown in the above figure, in the short column of high-socket spread footings, the steel grade of longitudinal bars is HRB400 and the steel grade of stirrups is HPB300. Reinforcement bars are arranged as corner bars (4 ϕ 20), middle bars (ϕ 16@ 220) on the longer side, and middle bars (ϕ 16@ 200) on the shorter side. The diameter of stirrups is 10. The spacing of stirrups in the socket of the short column is 150 and the spacing of stirrups in other parts in the short column is 300 (Only longitudinal bars and rectangular stirrups in the short column of the foundation are shown in the above figure)
4	**Elevation of the base of the foundation (optional)** (×.×××)	See Table 4.4.2
5	**Text annotations if necessary (optional)**	See Table 4.4.2

Specific annotations of normal spread footings   Table 4.4.4

Serial numbers	Denotation contents	Descriptions
1	**Dimensions of spread footing** x, y, x_c, y_c (or d_c of circular columns), $x_i, y_i, i=1,2,3\cdots\cdots$	x and y are side lengths of normal spread footings in two directions; y_c are sectional dimensions of the column; x_i and y_i are step widths or plane dimensions of the slope (when a short column exists, sectional dimensions of the short column should also be denoted). Specific annotations of normal spread footings with a symmetric stepped section are provided below. 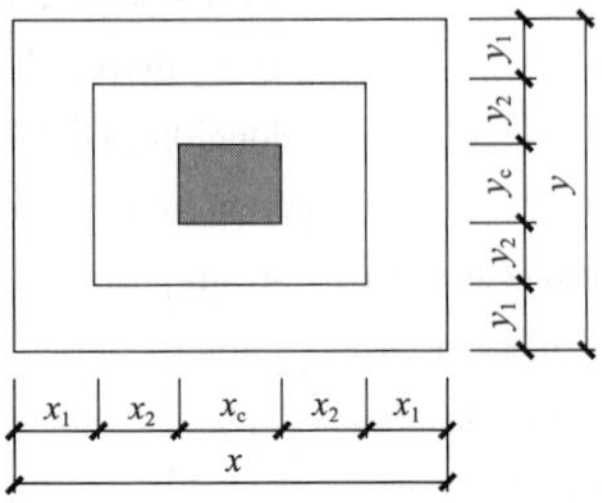 Specific annotations of normal spread footings with an asymmetric stepped section are provided below. 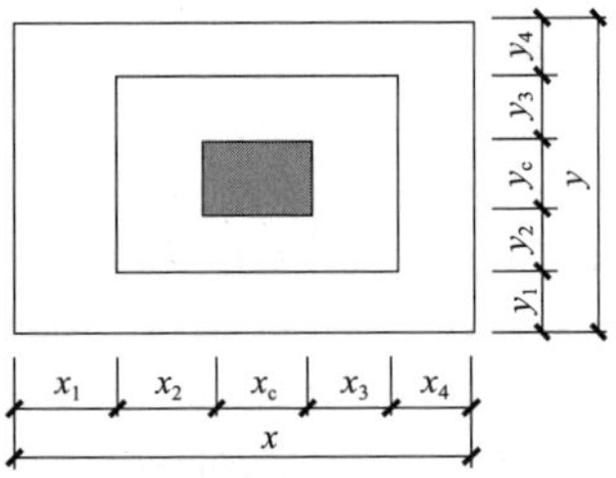 Specific annotations of normal spread footings with a symmetric sloped section are provided below. 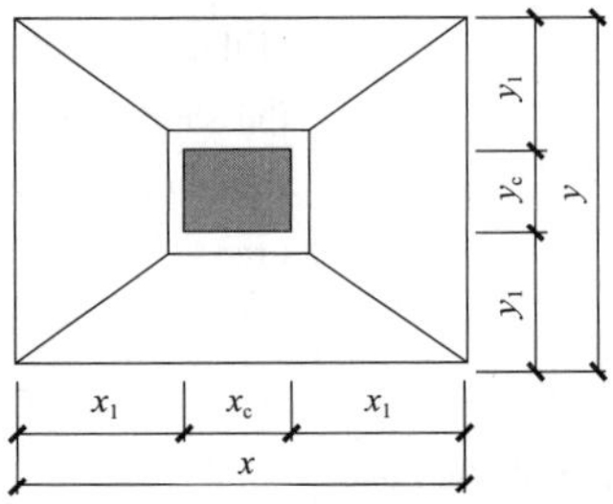 Specific annotations of normal spread footings with an asymmetric sloped section are provided below 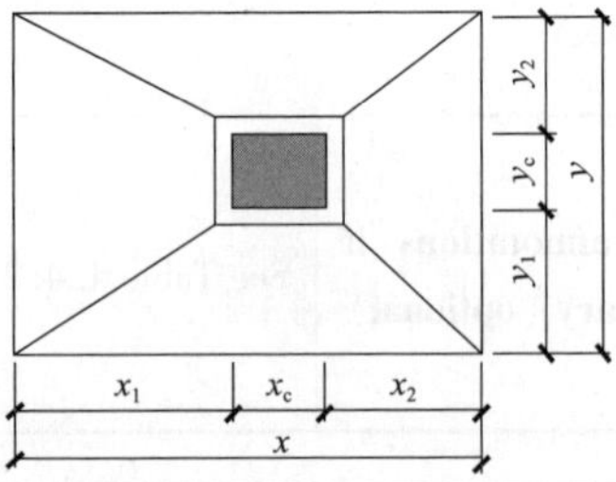

Specific annotations of socket spread footings   Table 4.4.5

Serial numbers	Denotation contents	Descriptions
1	**Dimensions of spread footing** $x, y, x_u, y_u, t_i, x_i,$ $y_i, i=1,2,3\cdots\cdots$	x and y are side lengths of socket spread footings in two directions; y_u are dimensions of the top of the socket; t_i is the thickness of the top cross-section of the socket wall and the thickness at the bottom of the socket wall is t_i+25; y_i are step widths or plane dimensions of the slope. Specific annotations of socket spread footings with a stepped section are provided below. 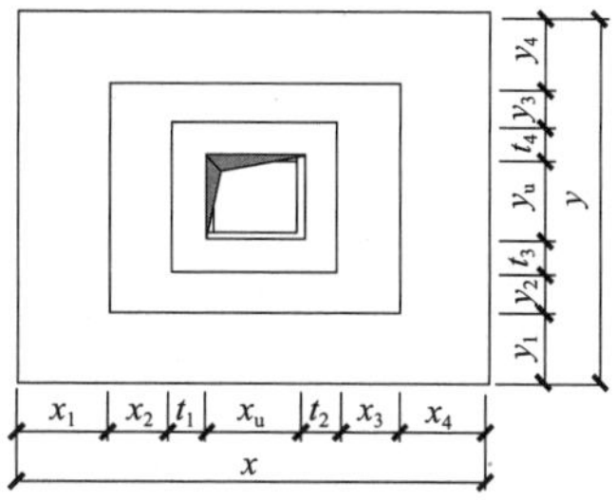 Specific annotations of socket spread footings with a sloped section are provided below. 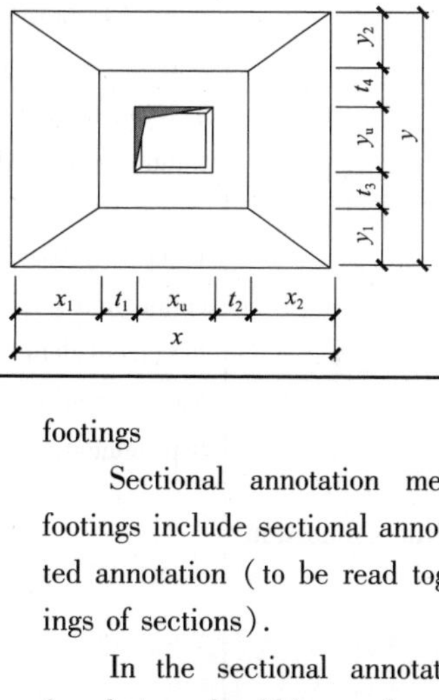

As for double-column spread footings with small spacing between two columns, in general, reinforcement bars are only required at the bottom of the foundation. If the column spacing is too large, reinforcement bars are required at the bottom and the top of the foundation or connection beams between two columns. As for four-column spread footings, two parallel connection beams may be arranged, and reinforcement bars may be placed at the top of the foundation between two connection beams.

Annotation methods of connection beams and reinforcement bars at the top of the foundation slab of double-column spread footings are shown in Table 4.4.6.

4) Sectional annotation method of spread footings

Sectional annotation methods of spread footings include sectional annotation and tabulated annotation (to be read together with drawings of sections).

In the sectional annotation method, all foundations should be numbered in the plan layout of the foundations. The numbering method is shown in Table 4.4.1.

(1) Sectional annotation

The format and contents of sectional annotations of a single foundation are the same as those of the traditional "orthographic projection representation method of a single member". When geometric dimensions have been shown clearly in the plan layout of foundations through specific annotation, it does not denote it again in sectional drawings. For details of a single foundation, standard detailing in this drawing collection may be taken as reference.

Annotation methods of double-column spread footings　　Table 4.4.6

Serial numbers	Denotation contents	Descriptions
1	**Bars at the top of the foundation slab** T: xx⌽xx@xx / ⌽xx@xx	Bars at the top of the foundation slab are usually placed at both sides of the centerline of two columns and are denoted as "longitudinal bars between two columns/distribution bars", marked with "T". If longitudinal main bars do not fully cover the top of the foundation slab, the total number of bars should be denoted. 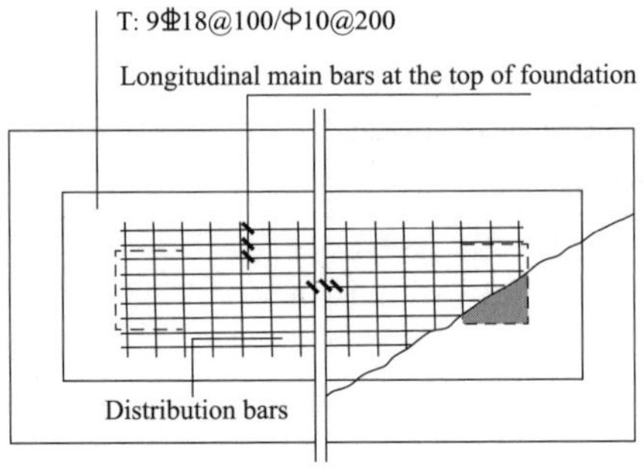 e.g. As shown in the above figure, there are 9 longitudinal main bars at the top of the spread footing. The steel grade is HRB400. The diameter is 18 and the spacing is 100. The steel grade of distribution bars is HPB300. The diameter is 10 and the spacing is 200
2	**Bars in connection beam**	If the double-column spread footing consists of foundation slabs and connection beams, beam numbers, geometric dimensions and reinforcement details of the connection beam should be denoted. The annotation method of the connection beam is the same as that in trip foundations 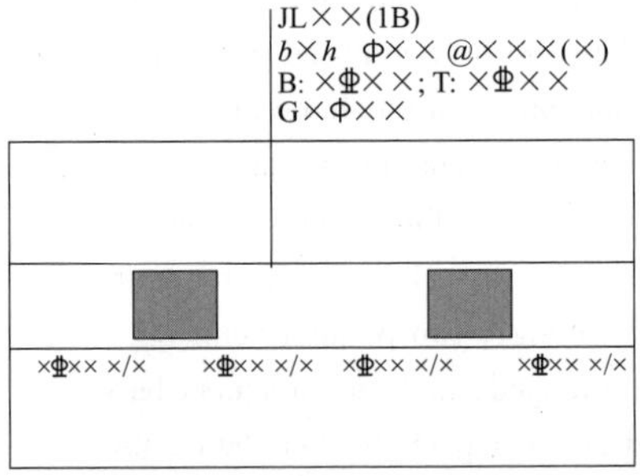
3	**Bars in a foundation slab** B: X ⌽xx@xx, Y ⌽xx@xx	Bars in a foundation slab may be denoted by the annotation method of those of foundation slabs of single-column spread footings. See Table 4.4.2

(2) Tabulated annotation

Tabulated annotation (to be read together with sectional drawings) may be used to summarise information of foundations with the same parameters. Geometric dimensions and reinforcement details of cross-sections should be provided in a table. Foundation numbers should be denoted in sectional drawings according to the corresponding items in Table 4.4.7.

Tabulated annotations of normal spread footings are shown in Table 4.4.7.

Tabulated annotations of geometrical dimensions and reinforcement details of normal spread footings
Table 4.4.7

Foundation number/ section informantion.	Geometric dimensions of sections				Reinforcement at the bottom (B)	
	x, y	x_c, y_c	x_i, y_i	$h_1/h_2/$	Direction X	Direction Y
DJ$_j$××					X ϕ××@××	Y ϕ××@××

5) An example of ichnographic construction drawing of spread footings

An example of ichnographic construction drawing of spread footing is shown in Figure 4.4.6

3. Ichnographic construction drawings of strip foundations

There are two kinds of strip foundations: strip foundation with beams and slab-type strip foundation, as shown in Figure 4.4.7.

Strip foundations with beams apply to reinforced concrete frame structures, frame-shear wall structures, partial frame-supported shear wall structures, and steel structures. In ichnographic construction drawings, beams and foundation slabs are usually represented separately.

Slab-type strip foundations apply to reinforced concrete shear wall structures and masonry structures. Only foundation slab details are shown in ichnographic construction drawings.

1) Representation method of ichnographic construction drawing of strip foundations

Annotation methods of ichnographic construction drawings of strip foundations include ichnographic annotation and sectional annotation. One or both of them can be used in the design of strip foundation construction drawings according to the specific situation of a project.

In the plan layout of strip foundations, the plan of strip foundations should be shown with columns and walls supported by foundations. If elevations of the bases of foundations are different, the scope and elevation differences relative to the reference elevation of the bases of foundations should be denoted.

If the centerline of the beam in strip foundations with beams or the centerline of slab-type strip foundation does not coincide with the gridline in architectural drawings, the distance should be denoted. If the foundation numbers of strip foundations are the same, it is only necessary to denote the foundation number on one strip foundation.

2) Numbering of strip foundations

The numbering of strip foundations includes numberings of foundation beams and foundation slabs, as shown in Table 4.4.8.

3) Ichnographic annotation method of foundation beams

Ichnographic annotation methods of foundation beams include general annotation and specific annotation. If the general annotation does not apply to a certain part of the foundation beams, its details should be denoted with a specific annotation. During construction, specific annotation has a higher priority.

(1) General annotation

General annotations of foundation beams include three compulsory items: foundation beam

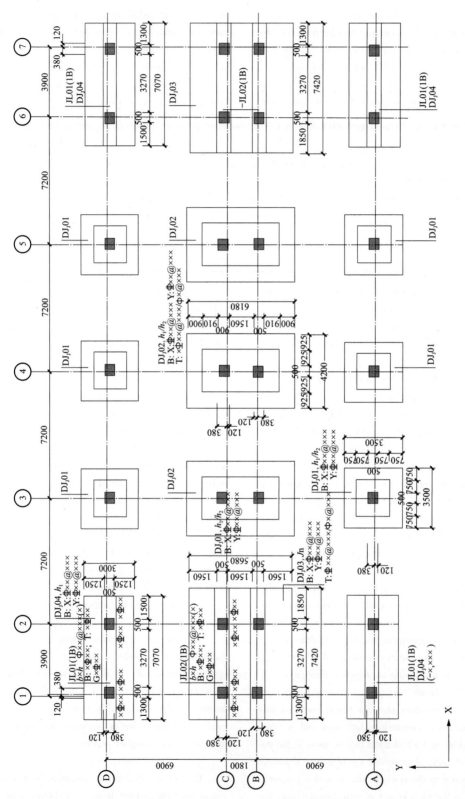

Figure 4.4.6　An example of construction drawing of normal spread footing using IRM

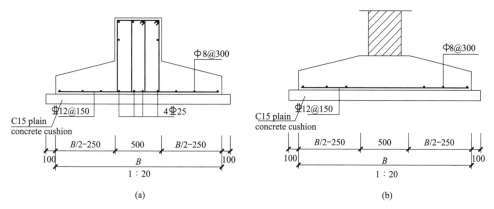

Figure 4.4.7 Strip foundations
(a) A strip foundation with beams;(b) A slab-type strip foundation

Numberings of beams and slabs of strip foundations   Table 4.4.8

Types		Type Codes	Serial numbers	Numbers of spans and cantilevers
Foundation beams		JL	××	(××) ends without cantilevers
Strip foundation slabs	Sloped	TJB_P	××	(××A) with a cantilever at one end
	Stepped	TJB_J	××	(××B) with cantilevers at both ends

Note: Sloped or single-step cross-sections are often used in strip foundations.

numbers, sectional dimensions, and reinforcement details. There are also two optional items: base elevation of foundation beams (if it is different from the reference elevation) and text annotations if necessary. General annotations of foundation beams are shown in Table 4.4.9.

General annotations of foundation beams   Table 4.4.9

Serial number	Denotation contents	Descriptions
1	Numbering of beams (compulsory) JL××(××A)	JL12(5A) See Table 4.4.8
2	Sectional dimensions of foundation beams (compulsory) $b \times h$	The width and depth of the section are denoted as $b \times h$. Haunched beams should be denoted as $b \times h\ Yc_1 \times c_2$, where c_1 indicates the length of the haunch and c_2 indicates the depth of the haunch
3	Reinforcement details (compulsory) Stirrups Φ××@××(××)	If there is only one spacing arrangement of stirrups, the steel grade, diameter, and spacing of stirrups and the number of legs should be denoted (the number of legs should be denoted in brackets)

Continued

Serial number	Denotation contents	Descriptions
3	(1) Stirrups ××⌽××@ ××/⌽ ××@ ××(××)	If there are two kinds of stirrups in a given design, "/" is used to separate different kinds of stirrups in the following way: the number, diameter, and spacing of stirrups arranged at two ends/diameter and spacing of stirrups arranged at midspan (it is not necessary to show the number of stirrups at midspan). e. g. "9⌽16@100/⌽16@200 (6)" indicates two different arrangements of stirrups in a foundation beam. In the vicinity of the ends of a foundation beam, the steel grade is HRB400. The diameter of stirrups is 16 and 9 stirrups are arranged with the spacing of 100. In the midspan of the foundation beam, the steel grade is HRB400. The diameter of stirrups is 16 and the spacing is 200. There are 6 legs in each stirrup
	(2) Longitudinal bars at the bottom B: ××⌽×× or B: ××⌽××+(××⌽××)	"B" is used to represent continuous longitudinal bars at the bottom of the beam (the total area of continuous bars should be no less than 1/3 of the total area of main bars at the bottom). If the number of continuous longitudinal bars at midspan is smaller than the number of legs of stirrups, erection bars should be placed at the bottom of the beam to fix the legs. It should be denoted as continuous longitudinal bars followed by "+" and erection bars in brackets
	(3) Longitudinal bars at the top T: ××⌽×× or T: ××⌽×× ×/×	"T" is used to represent continuous longitudinal bars at the top of the beam and bottom continuous longitudinal bars are separated by ";". If continuous longitudinal bar details in specific spans of the foundation beam are different from the general annotation, they should be denoted with a specific annotation. If there is more than one layer of continuous longitudinal bars at the bottom or the top of the beam, "/" is used to separate longitudinal bars in each layer from upper layers to lower layers. e. g. "B: 4⌽25; T: 12⌽25 7/5" indicates that 4⌽25 continuous longitudinal bars are placed at the bottom of the beam. 7⌽25 continuous longitudinal bars are placed in the upper layer at the top of the beam and 5⌽25 are placed in the lower layer. The total number of continuous longitudinal bars at the top is 12⌽25
	(4) Side bars G ××⌽×× or N ××⌽××	If the web thickness of the beam h_w is no less than 450, side bars should be placed according to requirements, prefix "G" is used to represent the total number of side bars placed symmetrically on both sides of the beam. e. g. G8⌽14 indicates that side bars placed on each side of the beam are 4⌽14 and that the total number of side bars is 8⌽14. Prefix "N" is used to represent torsion bars on both sides of the beam. e. g. N8⌽16 indicates that torsion bars are placed on both sides of the beam and that the total number is 8⌽16. The bars should be placed symmetrically and uniformly along with the depth of the section

Continued

Serial number	Denotation contents	Descriptions
4	Elevation of the base of foundation beams (optional) (x.xxx)	If the elevation of the base of strip foundations is different from the reference elevation, it should be denoted in brackets
5	Text annotations if necessary (optional)	If there are special requirements for foundation beams in the design, it is recommended to provide additional notes

(2) Specific annotation

Specific annotations of foundation beams include bottom longitudinal bars (all continuous longitudinal bars and discontinuous longitudinal bars) in the support of the foundation beam, additional stirrups/hangers, the depth of variable sections in the cantilever part of foundation beams, and revised information of specific annotation. Specific annotation of foundation beams is shown in Table 4.4.10.

Specific annotations of foundation beams　　　　Table 4.4.10

Serial numbers	Denotation contents	Descriptions
1	Bottom longitudinal bars in the support of a foundation beam xx⌽xx　x/x	It includes all continuous longitudinal bars and discontinuous longitudinal bars: (1) If there is more than one layer of longitudinal bars at the bottom, "/" should be used to separate longitudinal bars in each layer from upper layers to lower layers. (2) If the longitudinal bars in the same layer have two different diameters, "+" should be used to separate bars with different diameters. (3) If longitudinal bars placed at the bottom of foundation beams at both sides of the interior supports are different, bar information should be denoted at both sides of support ends. If bottom longitudinal bars placed at both sides of support ends are the same, only one end should be denoted. (4) If all bottom longitudinal bars in the beam at support ends are the same as bottom continuous longitudinal bars denoted in general annotation, specific annotation is not necessary. (5) Prefix "Y" in brackets is used to represent reinforcement bars placed at the haunched support of vertical-haunched beams. e.g. If Y4 ⌽ 25 is denoted at the end of the vertical-haunched beam (support), it indicates that inclined longitudinal bars in vertical haunch are 4 ⌽ 25

Continued

Serial numbers	Denotation contents	Descriptions
2	**Additional stirrups or hangers** ××⌽×× ××⌽××(××)	If there is no column sitting on the crossing joint of two foundation beams, additional stirrups or hangers should be placed if necessary. Additional stirrups or hangers should be drawn directly in the primary foundation beams of strip foundations in the plan and the total amount of reinforcement bars should be denoted with specific annotation (the number of stirrup legs should be denoted in brackets). If most of the stirrups or hangers are the same, they may be denoted in the same way. Those different from the most same stirrups or hangers should be denoted separately
3	**Depth of variable sections in the cantilever part of foundation beams**	$b \times h_1 / h_2$ should be denoted by specific annotation, where h_1 indicates the depth of section at the support end and h_2 indicates the depth of section at the free end
4	**Revised information on a specific annotation**	If some items in general annotations of the foundation beam (such as sectional dimensions, stirrups, continuous longitudinal bars at the bottom and the top of erection bars, side bars, and bottom elevations of foundation beam) do not apply to some spans or cantilevers, such details should be revised and denoted with a specific annotation. During construction, specific annotation has a higher priority

(3) An example of the ichnographic annotation method of foundation beams

Examples of the ichnographic annotation method of foundation beams are shown in Table 4.4.11.

4) Ichnographic annotation method of strip foundation slabs

Ichnographic annotation methods of strip foundation slabs include general annotation and specific annotation.

(1) General annotation

General annotations of strip foundation slabs include three compulsory items (slab numbers, vertical dimensions of cross-section, and reinforcement details) and two optional items [elevation of the base of strip foundation slabs (if it is different from the reference elevation) and text annotations if necessary]. General annotations of strip foundation slabs are shown in Table 4.4.12.

(2) Specific annotation

Specific annotations of strip foundation slabs include plane dimensions of the strip foundation slab and the revised information. Specific annotations of strip foundation slabs are shown in Table 4.4.13.

Examples of ichnographic annotation method of foundation beams

Table 4.4.11

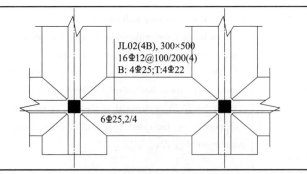

Continued

Methods	Denotation contents	Illustrations
General annotation	JL02 (4B)	**Numbering of foundation beam:** No. 2 foundation beam with four spans and cantilevers at both ends
	300×500	**Sectional dimensions of foundation beams** The width of the beam is 300mm and the depth is 500mm
	16 ⏀ 12 @ 100/200(4)	**Stirrup** The steel grade is HRB400. The diameter of stirrups is 12 and 16 stirrups are arranged with a spacing of 100. In the midspan of the foundation beam, the steel grade is HRB400; the diameter of stirrups is 12; the spacing is 200. There are 4 legs in each stirrup
	B:4 ⏀ 25; T: 4 ⏀ 22	**Longitudinal bars at the bottom;** **Longitudinal bars at the top** 4 ⏀ 25 continuous longitudinal bars are placed at the bottom of the beam; 4 ⏀ 22 continuous longitudinal bars are placed at the top of the beam. The steel grade is HRB400. The diameter of longitudinal bars at the bottom is 25 and the diameter of longitudinal bars at the top is 22
Specific annotation	6 ⏀ 25 2/4	**Bottom longitudinal bars in the support** 2 ⏀ 25 continuous longitudinal bars are placed in the upper layer at the bottom of the beam and 4 ⏀ 25 are placed in the lower layer. The total number of continuous longitudinal bars at the bottom is 6 ⏀ 25. The steel grade is HRB400 and the diameter of longitudinal bars at the bottom is 25

General annotations of strip foundation slabs Table 4.4.12

Serial numbers	Denotation contents	Descriptions
1	**Numbering of slabs (compulsory)** $TJB_p \times \times (\times \times)$	$TJB_p06(6B)$ See Table 4.4.8
2	**Vertical dimensions of cross-section (compulsory)** $h_1/h_2/\cdots\cdots$	(1) Slabs with a sloped cross-section. 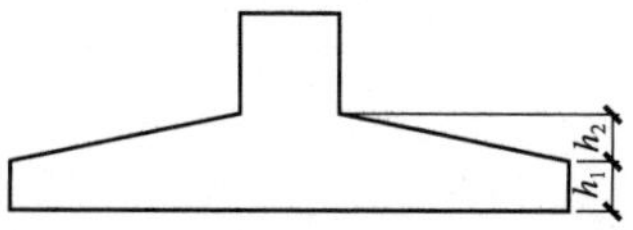

Continued

Serial numbers	Denotation contents	Descriptions
2	**Vertical dimensions of cross-section** (**compulsory**) $h_1/h_2/\cdots\cdots$	**e. g.** If vertical dimensions of the cross-section are denoted as 350/250, it indicates $h_1 = 350$ and $h_2 = 250$. The total depth of the strip foundation slab is 600. (2) Slabs with a stepped cross-section. 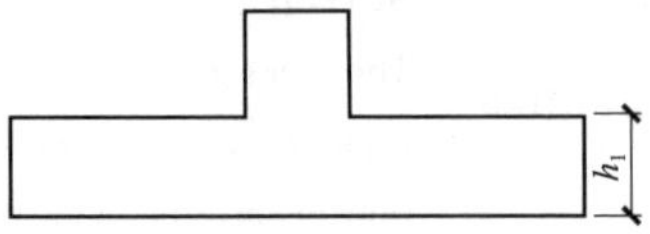 **e. g.** If vertical dimensions of the stepped cross-section are denoted as 400, it indicates $h_1 = 400$ and the total depth of the slab is 400
3	**Reinforcement details at the bottom and the top of the strip foundation slab** (**compulsory**) reinforcement details of foundation slabs B:⏀××@×××/⏀××@××× or B:⏀××@×××/⏀××@××× T:⏀××@×××/⏀××@×××	Transverse main bars at the bottom of the slab should be denoted with "B". The main transverse bars at the top of the slab should be denoted with "T". "/" should be used to separate the main transverse bars from longitudinal distribution bars in the strip foundation slab. 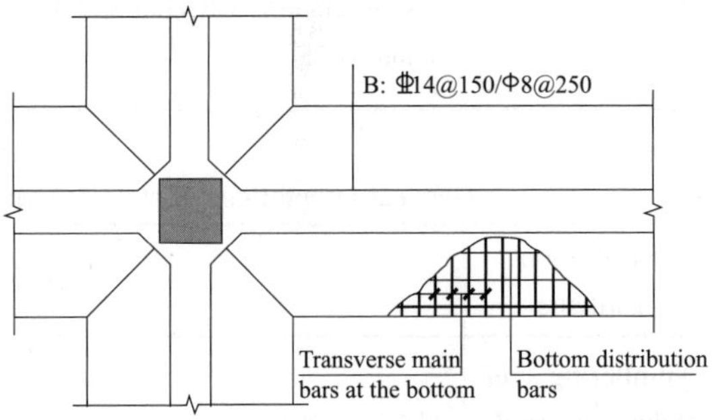 B: ⏀14@150/Φ8@250 Transverse main bars at the bottom Bottom distribution bars e. g. As shown in the above figure, the main transverse bars with the steel grade of HRB400 are placed at the bottom of the slab. The diameter is 14 and the spacing is 150. There are also longitudinal distribution bars with the steel grade of HPB300 placed at the bottom of the slab. The diameter is 8 and the spacing is 250.

Continued

Serial numbers	Denotation contents	Descriptions
3	**Reinforcement details at the bottom and the top of the strip foundationslab** (**compulsory**) reinforcement details of foundation slabs B:⌽××@×××/⌽××@××× or B:⌽××@×××/⌽××@××× T:⌽××@×××/⌽××@×××	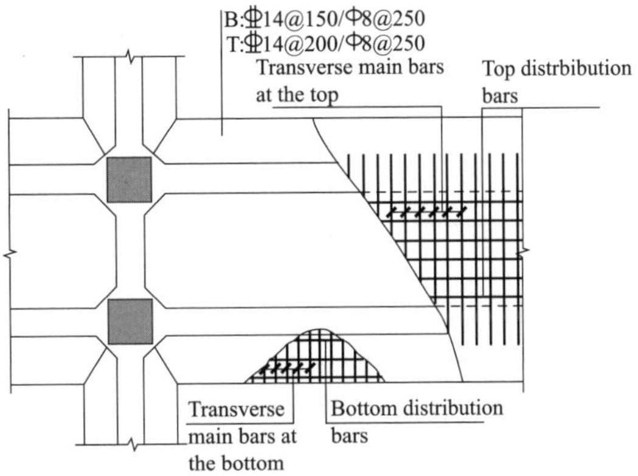 B:⌽14@150/⌽8@250 T:⌽14@200/⌽8@250 Transverse main bars at the top — Top distribution bars Transverse main bars at the bottom — Bottom distribution bars e.g. As shown in the above figure, as for the strip foundation slab with double beams or double walls, besides the bars placed at the bottom of the slab, it is necessary to place bars between two beams or two walls at the top of the slab
4	**Elevation of the base of strip foundation slabs** (**optional**) (x.×××)	If the elevation of the base of trip foundation slabs is different from the reference elevation, it should be denoted in brackets.
5	**Text annotations if necessary** (**optional**)	If there are special requirements for strip foundation slabs in the design, necessary text annotations should be added

Specific annotations of strip foundation slabs   Table 4.4.13

Serial numbers	Denotation contents	Descriptions
1	**Plane dimensions of the strip foundation slabs** b, b_i $i=1, 2 \cdots$	Plane dimensions of the strip foundation slab are denoted with specific annotation as b and b_i, where $i=1, 2, \cdots$ b indicates the total width of the foundation slab and b_i indicates the width of steps in foundation slabs. For the strip foundation slab with a symmetric sloped or single-step cross-section in respect to the foundation beam, b_i may be omitted Only one slab should be denoted if the numbers of strip foundation slabs are the same

Continued

Serial numbers	Denotation contents	Descriptions
2	Revised information	If the details of general annotation on the strip foundation slab, such as vertical dimensions of slab cross-section, reinforcement details of the slab, and a base elevation of the slab, do not apply to some spans or cantilever parts of the strip foundation slab, the revised information may be provided through specific annotation. During construction, specific annotation has a higher priority

(3) Example of ichnographic annotation method of strip foundation slabs

An example of the ichnographic annotation method of strip foundation slabs is shown in Table 4.4.14.

5) Sectional annotation methods of strip foundations

Sectional annotation methods of strip foundations include sectional annotation and tabulated annotation (to be read together with sectional drawings).

In the sectional annotation method, all strip foundations should be numbered in the plan layout of the foundations. The numbering method is shown in Table 4.4.8.

(1) Sectional annotation

Example of ichnographic annotation method of strip foundation slabs

Table 4.4.14

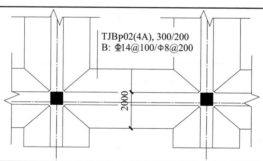

Methods	Denotation contents	Illustrations
General annotation	TJB$_p$02 (4A)	**Numbering of slabs** The No. 2 strip foundation has two spans and a cantilever at one end
	300/200	**Sectional dimensions of slabs** As for slabs with sloped cross-section, it indicates $h_1 = 300$ and $h_2 = 200$ (the total depth of the strip foundation slab is 500)
	B: ⌀14@100/ ⌀8@200	**Transverse main bars at the bottom and longitudinal distribution bars** It is indicated that main transverse bars with the steel grade of HRB400 are placed at the bottom of the slab. The diameter is 14 and the spacing is 100. There are also longitudinal distribution bars with the steel grade of HPB300 placed at the bottom of the slab. The diameter is 8 and the spacing is 200

Continued

Methods	Denotation contents	Illustrations
Specific annotation	2000	**Plane dimensions of the strip foundation slab** The total width of the foundation slab is 2000

The format and items of sectional annotations of strip foundations are the same as those of the traditional "orthographic projection representation method of a single member". When geometric dimensions have been shown clearly in the plan layout of foundations with specific annotation, it is not necessary to denote them again in sectional drawings.

(2) Tabulated annotation

Tabulated annotation (to be read together with sectional drawings) may be used to summarise the information of strip foundations with the same parameters. Geometric dimensions and reinforcement details of cross-sections should be provided in a table. Foundation numbers should be denoted in sectional drawings. Tabulated annotations of foundation beams are shown in Table 4.4.15. Tabulated annotations of foundation slabs are shown in Table 4.4.16.

6) Example of ichnographic construction drawing of strip foundations

Table of geometric dimensions and reinforcement details of foundation beams

Table 4.4.15

Foundation beam number/ section number	Geometric dimensions of the cross-section		Reinforcement details	
	$b \times h$	Vertical haunched $c_1 \times c_2$	Continuous longitudinal bars at the bottom + discontinuous longitudinal bars, continuous longitudinal bars at the top	First stirrups/second stirrups
JL××(××A)	$b \times h$	Y$c_1 \times c_2$	B:××⌽×× +(××⌽××); T:××⌽ ××	××⌽××@ ××/⌽×× @ ××(××)

Note: If necessary, additional information may be added according to actual situations. For example, if the elevation of the base of the foundation is not the same as the reference elevation, it should be denoted.

Table of geometric dimensions and reinforcement details of the slabs of strip foundations

Table 4.4.16

Foundation beam number/ section number	Sectional dimensions			Reinforcement at the bottom (B)	
	b	b_i	h_1/h_2	Transverse main bars	Longitudinal distribution bars
TJB$_p$××(××)				B:⌽××@ ×××	⌽××@ ×××

Note: If necessary, additional information may be added according to actual situations. For example, information on additional bars at the top of the foundation beam and the elevation of the base of the foundation may be added if it is not the same as the reference elevation.

An example of ichnographic construction drawing of strip foundations is shown in Figure 4.4.8.

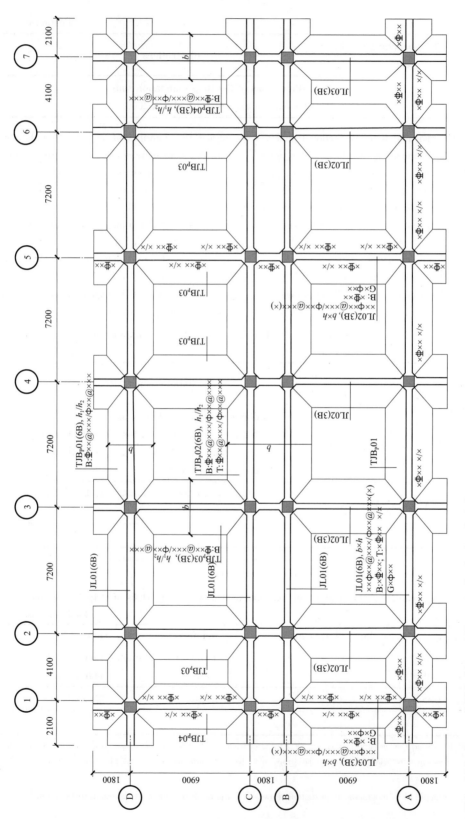

Figure 4.4.8 An example of the ichnographic construction drawing of strip foundations

4.5 Ichnographic Construction Drawings of the Main Structure

The main structure is the structure above the foundation. It bears all the building loads and transfers the loads to the foundation. According to the structural system, the cast-in-situ reinforced concrete structures can be classified into frame structure, shear wall structure, frame-shear wall structure, and tube structure. These structures are composed of several vertical and horizontal members. The common vertical members include columns and shear walls. The common horizontal members include beams and slabs. This section mainly introduces cast-in-situ columns, beams, and slabs.

1. Ichnographic construction drawings of columns

Columns are main compression members bearing vertical loads. Common columns include rectangular columns, circular columns, and core columns. The bottom of the columns is directly supported by the foundation and beams or shear walls. This subsection mainly introduces ichnographic construction drawings of columns.

1) Representation method of ichnographic construction drawing of columns

Ichnographic construction drawings of columns are to show column information with **tabulated annotation** method or **sectional annotation** method in the plan layout of columns at a proper scale.

In ichnographic construction drawings of columns, structural floor level, structural story height, and corresponding floor numbers shall be denoted. The position of the fixed end of the superstructure shall also be denoted.

The position of the fixed end of the superstructure:

(1) No note is required if the position of the fixed end of frame columns is on the top of the foundations.

(2) When the position of the fixed end of frame columns is not on the top of the foundations, the level of the fixed end shall be indicated with double fine underlines in the floor level table and the level of the fixed end of the superstructure shall be particularly noted under the table.

(3) If the position of the fixed end of frame columns is not on the top slab of the basement and a fixing contribution of the top slab of the basement to the superstructure has to be considered, the level of the basement top slab can be indicated with double dash underlines.

2) Column number

A column number is composed of type code and serial number, as shown in Table 4.5.1.

Column Number

Table 4.5.1

Column types	Type codes	Serial numbers
Frame columns	KZ	××
Transfer columns	ZHZ	××
Core columns	XZ	××
Columns supported by beams	LZ	××
Columns supported by shear walls	QZ	××

Note: The same column number can be used if the height, dimensions of each section, and reinforcement details are the same. If only the sectional dimensions relative to the gridline in various columns are different, these columns may be denoted with the same number, but the sectional dimensions relative to the gridline must be denoted in the drawing.

3) Tabulated annotation method

The tabulated annotation method is to represent one or some typical columns (frame columns, transfer columns, columns supported by beams, and columns supported by shear walls) in the plan layout of columns. One typical column is selected from the columns with the same column number and denoted with the type code of geometric parameters. All details, such as column number, level of the starting and ending points of the column, geometric dimensions (including sectional dimensions of columns) and the reinforcement details shall be shown in a table. Shapes of column sections and details of stirrups shall also be denoted in ichnographic construction drawings of columns.

The contents in the table of columns are introduced below:

(1) Column number

Column number shall comply with Table 4.5.1.

(2) Levels of the starting and ending points of each section of columns

Levels of the starting and ending points of each section from the bottom of the column upwards shall be denoted. The column shall be divided into different sections if the cross-section or reinforcement bars are changed.

The level of the bottom of the frame column and transfer column is the level of the top of the foundation. The level of the bottom of the core column is the actual level of the starting point of the core column. The level of the bottom of columns supported by beams is the level of the top of beams. The level of the bottom of columns supported by shear walls is the level of the top of shear walls.

(3) Geometric dimensions

As for rectangular columns, sectional dimension $b \times h$, as well as geometric parameters b_1, b_2, h_1, and h_2 which are sectional dimensions relative to the gridline, shall be denoted for each section of the column ($b = b_1 + b_2$, $h = h_1 + h_2$). If one side of the section is overlapped with the gridline or located at the other side of the gridline, b_1, b_2, h_1 or h_2 may be zero or negative.

As for circular columns, $b \times h$ in the table of columns shall be replaced by the diameter of the column with the prefix "d". Simply, the relative position of the circular column section to the gridline is also represented by b_1, b_2, h_1, and h_2, where $b = b_1 + b_2 = h_1 + h_2$.

Core columns can be arranged in the center of some frame columns (column number shall be denoted) within a certain height based on the structural design. Core columns, of which the center coincides with that of columns, shall be noted with sectional dimensions. The setting out point of the core column is the same as that of the corresponding frame column. No independent sectional dimensions for the core column are required.

(4) Longitudinal bars in columns

If the diameters of all longitudinal bars are the same and the number of bars on each side of columns (including rectangular columns, circular columns, and core columns) are the same, information of longitudinal bars shall be denoted in "Total longitudinal bars" in the table of columns. In other cases, longitudinal bars in columns shall be respectively denoted as corner bars, middle bars on side b, and middle bars on side h. If bars are arranged symmetrically in a rectangular column, only middle bars on one side shall be denoted. Otherwise, middle bars on each side must be denoted.

(5) Stirrup type and number of stirrup legs

They shall be indicated in the corresponding column of the table. The shape and overlapping details of stirrups shall be shown in the upper lines in the column table or at a proper location of the drawing, with b, h, and stirrup type information corresponding to the column table.

(6) Steel grade, diameter, and spacing of column stirrups

Column stirrups (includes steel grade, diameter, and spacing) shall be denoted. "/" is used to distinguish different spacings of stirrups in stirrup densification zones at the ends of the column and other zones. Contractors shall adopt the maximum length of stirrup densification zones according to the requirements in the standard detailing drawings. If the stirrups in the frame joint zone are different from those in stirrup densification zones, the diameter and spacing of stirrups in the joint zone shall be indicated in brackets.

E. g. $\phi 10@100/200$ indicates that the steel grade of stirrups is HPB300 and that the diameter is 10. The spacing is 100 in stirrup densification zones and 150 in other zones.

$\phi 10@100/200$ ($\phi 12@100$). The stirrup information in brackets indicates that in the frame joint zone, the diameter and spacing of stirrups are respectively 12 and 100 and that the steel grade is HPB300.

If the spacing of stirrups along the whole column keeps constant, no "/" is required.

E. g. $\phi 10@100$ indicates that the stirrups are arranged along with the whole column height with a constant spacing of 100. The diameter is 10 and the steel grade is HPB300.

When spiral stirrups are adopted for circular columns, stirrups shall be described with the prefix "L".

E. g. $\phi L10@100/200$ indicates that spiral stirrups are arranged along with the column height with a spacing of 100 in stirrup densification zones and a spacing of 200 in other zones. The diameter is 10 and the steel grade is HPB300.

An example of ichnographic construction drawing of columns with tabulated annotation is shown in Figure 4. 5. 1.

4) Sectional annotation method

Sectional annotation method is to represent ichnographic construction drawing of columns with one or some typical columns in the plan layout of columns. One typical column is selected from the columns with the same column number and denoted with sectional dimensions and reinforcement details.

Except for core columns, all columns shall have different column numbers according to related requirements. The section of a typical column, which can represent the columns with the same column number, shall be drawn at a larger scale to show reinforcement details. Other individual information of each typical column, such as sectional dimensions $b \times h$, corner bars or total longitudinal bars (when all longitudinal bars in a section are the same and can be shown clearly in the drawing), amount of stirrups and the sectional dimensions b_1, b_2, h_1, and h_2, shall be added after the column number.

If the diameter of middle longitudinal bars is different from that of corner bars, additional information of the middle bars shall be added. As for a rectangular column section with symmetrical reinforcement, the annotation on one side is enough.

When the core column is designed within a certain height at the center of some frame columns, the core column shall be numbered. The level of the starting and ending points of the core column shall be provided after the column number and the total number of longitudinal bars and stirrups. If sectional dimensions of core columns are determined by detailing requirements and will be constructed according to standard detailing drawings, no special annotation for core columns is required. Core column position shall be arranged after the corresponding frame column. The sectional dimensions for core columns are not required.

In the sectional annotation method, for the columns with the same sectional dimensions and reinforcement details but different relative positions to the gridline, the same column number can be used. However, the sectional dimensions relative to the gridline shall be denoted in the sections without reinforcement details.

An example of ichnographic construction drawings of columns with **sectional annotation** is shown in Figure 4. 5. 2.

2. Ichnographic construction drawings of beams

The beams are horizontally placed flexural members. The beams bear the loads from slabs and transfer the loads to columns. According to types of supports, beams may be classified into simply supported beams, continuous beams, and cantilever beams. The reinforcement bars in beams mainly include bottom longitudinal bars, top longitudinal bars, side bars, stirrups, etc.

1) Representation method of ichnographic

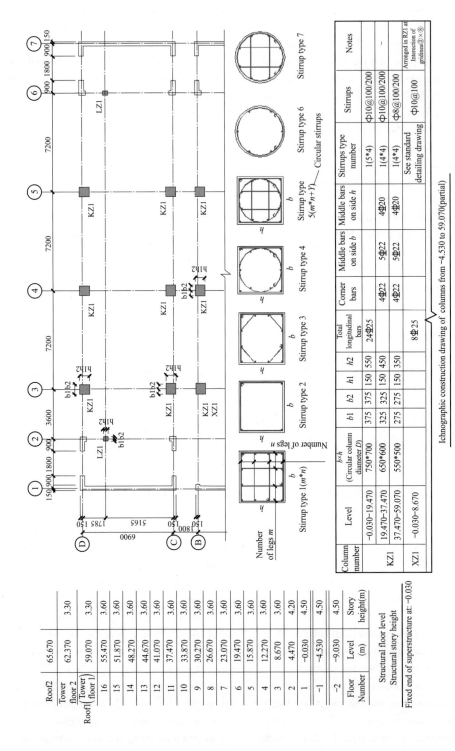

Figure 4.5.1 An example of the ichnographic construction drawing of columns with tabulated annotation

Chapter 4 Structural Construction Drawings

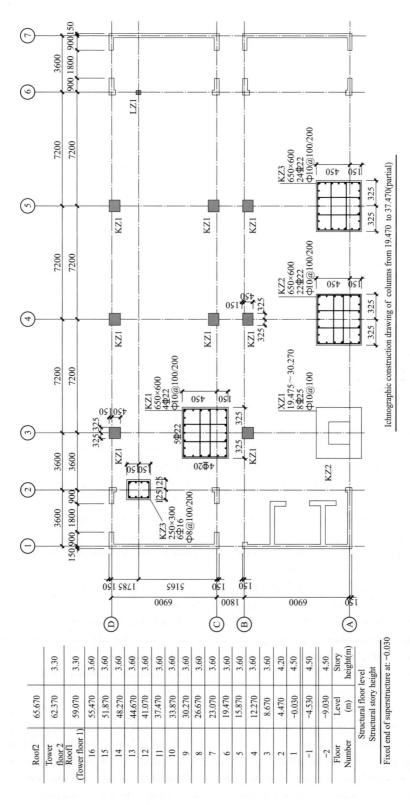

Figure 4.5.2 An example of the ichnographic construction drawing of columns with sectional annotation

construction drawings of beams

Ichnographic construction drawings of beams are to show beam information with the **ichnographic annotation method** or **sectional annotation method** in the plan layout of beams.

The plan layout of beams is to show beams and related columns, walls, slabs on the same structural floor at a proper scale. In ichnographic construction drawings of beams, the structural floor level, and corresponding floor number shall be denoted. If the beam axis does not coincide with the gridline, sectional dimensions shall be denoted. If the beam face coincides with the column face, no annotation is required for the beam.

2) Beam number

A beam number is composed of type code, serial number, number of spans, and cantilever information, as shown in Table 4.5.2.

Beam number   Table 4.5.2

Beam types	Type codes	Serial numbers	Number of spans and cantilever information
Floor frame beams	KL	××	(××), (×× A) or (×× B)
Floor frame flat beams	KBL	××	(××), (×× A) or (×× B)
Roof frame beams	WKL	××	(××), (×× A) or (×× B)
Frame-supported beams	KZL	××	(××), (×× A) or (×× B)
Column supporting transfer beams	TZL	××	(××), (×× A) or (×× B)
Non-frame beams	L	××	(××), (×× A) or (×× B)
Cantilever beam	XL	××	(××), (×× A) or (×× B)
#-shape arranged beams	JZL	××	(××), (×× A) or (×× B)

Note: (××A) indicates cantilever beam at one end; (××B) indicates cantilever beams at both ends; cantilevers are not counted in the number of spans.

3) Ichnographic annotation method

The ichnographic annotation method is to represent sectional dimensions and reinforcement information on typical beams in the plan layout of beams. Each typical beam represents the beams with the same beam number.

Ichnographic annotations of beams include two parts: general annotation and specific annotation for special parameters of beams. If the information in general annotations does not apply to some parts of beams, the specific annotation may be adopted. Information in the specific annotation has a higher priority during construction, as shown in Figure 4.5.3.

(1) General annotation

General annotations of a beam generally contain five compulsory items and one optional item and can be added on any span of a beam. General annotation is introduced below.

The five compulsory items are beam number, sectional dimensions, stirrup information, top continuous bars (or erection bars), and side bars (or torsion bars). The optional item is the level difference of beam top. General annotations of beams are shown in Table 4.5.3.

(2) Specific annotation

Specific annotations include special parameters of beams. Information in the specific annotation has a higher priority in the construction process. Specific annotations of beams are shown in Table 4.5.4.

Figure 4.5.4 shows an example of ichnographic construction drawing of beams with ichnographic annotation.

4) Sectional annotation method

Sectional annotation method is to represent one or some typical beams, which can represent the beams with the same beam number, in the plan layout of beams of each typical floor. Sectional dimensions and reinforcement details shall be denoted in the sectional annotation drawings indicated with the section number.

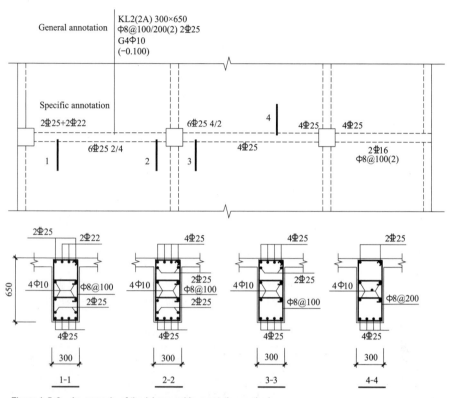

Figure 4.5.3 An example of the ichnographic annotation method
Note: Compared with IRM, these four beam sections are represented with the traditional method to show the same information. However, in ichnographic construction drawings, reinforcement details of the sections or cross-section numbers shown in Figure 4.5.3 are not necessary.

General annotations of beams   Table 4.5.3

Serial numbers	Denotation contents	Descriptions
1	**Beam number** (**compulsory**) KL××(××A)	KL12(5A) See Table 4.5.2
2	**Sectional dimensions of beams** (**compulsory**) $b×h$ $b×h\ Yc_1×c_2$	For beams with the same cross-section, the width and depth of the beams are denoted as $b×h$ For beams with a vertical haunch, the beam section is denoted as $b×h\ Yc_1×c_2$, where c_1 is the length of the haunch and c_2 is the depth of the haunch

Continued

Serial numbers	Denotation contents	Descriptions
2	$b \times h\ \mathrm{PY}c_1 \times c_2$	For beams with a horizontal haunch at one side, the beam section is denoted as $b \times h\ \mathrm{PY}c_1 \times c_2$, where c_1 and c_2 are respectively the length and width of the haunch. The haunch shall be denoted in the plan layout 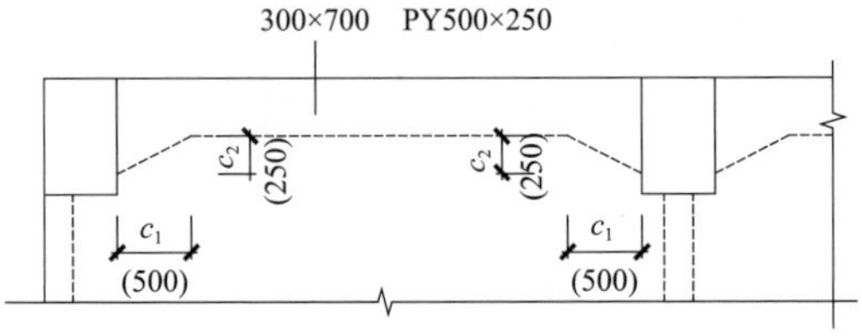
	$b \times h_1/h_2$	For cantilever beams with different depths at supported and free ends, a slash is used to distinguish the difference as follows: $b \times h_1/h_2$ 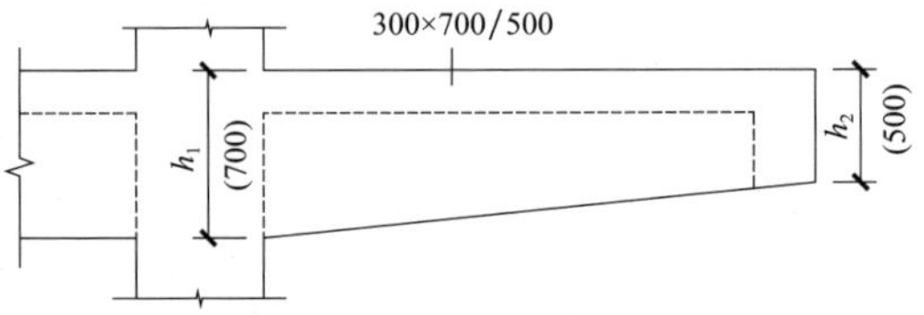
3	**Stirrup information (compulsory)** $\phi\ \times\times@\times\times/\times\times(\times\times)$ or ϕ $\times\times@\times\times(\times\times)/\times\times(\times\times)$	Stirrup information includes steel grade, diameter, and spacing of stirrups, and the number of legs in stirrup densification zones and other zones. Different stirrup spacing and the number of legs in stirrup densification zones and other zones shall be distinguished by "/". If the stirrup spacing and the number of legs are constant, no "/" is required. If the number of legs in stirrups along the beam is constant, it may be denoted only once. The number of legs shall be denoted in brackets. The length of stirrup densification zones may refer to standard detailing drawings of corresponding seismic grade. e.g. $\phi10@100/200(4)$ indicates that the steel grade of stirrups is HPB300 and that the diameter is 10. The spacings in stirrup densification zones and other zones are 100 and 200, respectively. The number of legs in the stirrup is 4 along the whole beam. $\phi8@100(4)/150(2)$ indicates that the steel grade of stirrups is HPB300 and that the diameter is 8. The spacings in stirrup densification zones and other zones are 100 and 150 respectively. The numbers of legs in stirrup densification zones and other zones are 4 and 2, respectively

Continued

Serial numbers	Denotation contents	Descriptions
4	Top continuous bars or erection bars (compulsory) ××⌽×× or (××⌽××)	Steel grade, diameter, and the number of bars are designed according to structural strength requirements and other detailed requirements such as the number of stirrup legs. If both continuous bars and erection bars are arranged in the same layer, "+" shall be used between continuous bars and erection bars as "information of corner longitudinal bars +(information of erection bars)" to distinguish bar types and diameters. If all top bars are erection bars, the bar information shall be denoted in brackets only. e. g. 2⌽22 indicates that stirrups with 2 legs are placed together with 2⌽22 continuous bars. 2⌽22+(4⌽12) indicates that stirrups with 6 legs are placed together with 2⌽22 continuous bars and 4⌽12 erection bars. When longitudinal bars placed at the top and bottom of the beam in a span are the same and the same arrangement keeps constant in most spans, information of bottom longitudinal bars can be added in the above information, ";" shall be used to distinguish top and bottom longitudinal bars. e. g. 3⌽22;3⌽20 indicates that top continuous bars in the beam are 3⌽22 and that bottom continuous bars in the beam are 3⌽20
5	Side bars or torsion bars on both sides of the beam (compulsory) G××⌽×× or N××⌽××	If the depth of beam web h_w >450, side bars are required and steel grade, diameter, and the number of side bars shall comply with the requirements in design codes. Side bars are denoted with the prefix "G" followed by the total number of symmetrically arranged bars. e. g. G4⌽12 indicates that a total of 4⌽12 side bars are arranged in the beam with 2⌽12 on each side. If longitudinal torsion reinforcement is necessary for beams, it shall be denoted with the prefix "N" followed by the number of symmetrically arranged torsion bars. Torsion bars shall meet the requirements for the spacing of side bars, If torsion bars are arranged, side bars are no longer required. e. g. N6⌽22 indicates that a total of 6⌽22 torsion bars are arranged in beam with 3⌽22 on each side
6	Level difference of beam top (optional)	The level difference of beam top is the level difference between the beam and the structural floor in which the beam is located. For beams in the mezzanine, it indicates the difference between the beam top and the mezzanine floor level. If the level difference exists, it shall be denoted in brackets. If there is no level difference, no note is required. If the beam top is above the structural floor in which the beam is located, the level difference shall be positive. Otherwise, the level difference shall be negative. e. g. When the levels of the typical structural floors are 44.950m and 48.250m and the level difference of the top of a beam top is denoted as (-0.050), it indicates that the level of the beam top is 0.05m below 44.950m and 48.250m

Specific annotations of beams   Table 4.5.4

Serial numbers	Denotation contents	Descriptions
1	Top longitudinal bars at the support ××⌀×× or ××⌀×× ××/×× or ××⌀×× + ××⌀××	All longitudinal bars including top continuous bars shall be denoted: (1) When more than one layer of longitudinal bars are arranged at the top of the beam, bar information of each layer shall be denoted from the top downwards and "/" is used to distinguish the layers. e. g. 6 ⌀ 25 4/2 indicates 4 ⌀ 25 in the upper layer and 2 ⌀ 25 in the lower layer of the top longitudinal bars at the supports of the beam. (2) If longitudinal bars with two different diameters are arranged in the same layer, the two diameters will be denoted with "+" between them and the corner bars shall be denoted first. e. g. If there are four top bars at the support, 2 ⌀ 25 in the corners and 2 ⌀ 22 in the middle, it shall be denoted as 2 ⌀ 25+2 ⌀ 22. (3) If top longitudinal bars in beams on both sides of the interior support are different, the information shall be denoted on both sides. If top longitudinal bars in beams on both sides of the interior support are the same, the information can be denoted at one side only
2	Bottom longitudinal bars in beams ××⌀×× or ××⌀×× ××/×× or ××⌀×× + ××⌀××	(1) When more than one layer of longitudinal bars are placed at the bottom, bar information of each layer shall be denoted from the top downwards and "/" is used to distinguish the layers. e. g. 6 ⌀ 25 2/4 indicates 2 ⌀ 25 longitudinal bars in the upper layer and 4 ⌀ 25 longitudinal bars in the lower layer, which all extend into the support. (2) If longitudinal bars with two different diameters are arranged in the same layer, the two diameters shall be denoted with "+" between them and the corner bars shall be denoted first. (3) If all bottom longitudinal bars do not extend into the support, the number of the bars cut off in the span shall be denoted in brackets. e. g. 6 ⌀ 25 2(-2)/4 indicates that there are two layers of bottom longitudinal bars in the beam, 2 ⌀ 25 in the upper layer and cut off in span, 4 ⌀ 25 in the lower layer and extending into the support. 2 ⌀ 25+3 ⌀ 22(-3)/5 ⌀ 25 indicates that there are two layers of bottom longitudinal bars in the beam, 2 ⌀ 25 and 3 ⌀ 22 in the upper layer in which 3 ⌀ 22 are cut off in the span, and 5 ⌀ 25 in the lower layer and extending into the support. (4) If the longitudinal bars at the top and bottom of the beam are continuous, they are denoted according to general annotation and no additional denotation is required in the specific annotation
3	Revised information of a general annotation	If general annotation (one or several items of sectional dimensions, stirrups, top continuous bars or erection bars, side bar or torsion bars, and a level difference of beam top) can not apply to some spans or cantilever part, information may be indicated in the specific annotation at the corresponding span or cantilever part. Such information shall be referred to during the construction process

Continued

Serial numbers	Denotation contents	Descriptions
4	Additional stirrups or hangers ××⌽×× (××) or ××⌽××	New stirrups or hangers shall be denoted in the primary beam directly. The total amount of bars shall be noted (number of legs of additional stirrups shall be denoted in brackets). If most of the additional stirrups or hangers are the same, the information may be denoted in ichnographic construction drawings. Other additional stirrups or hangers with different amounts shall be denoted in the specific annotation

All beams shall be numbered according to Table 4.5.2. For each typical beam, section number shall be denoted on one side of the beam, and reinforcement details of the section shall be denoted in the same drawing or other drawings. If the top level of a beam is different from the level of the structural floor, the level difference of the beam top shall be added after the beam number.

In the sectional annotation method, sectional dimensions $b \times h$ and the amounts of top and bottom bars, side bars, or torsion bars shall be denoted in the same way as the ichnographic annotation method.

As for frame flat beams, the number of main bars which do not pass through the column section shall be denoted in the detailing drawing of the section. As for additional bars in frame flat beam-column joints, the details of additional longitudinal bars in frame flat beam-column joints, all vertical tie bars in the core region outside column, and additional U-shaped stirrups at end supports shall be indicated in the plan and cross-section drawings. The number of bars shall be denoted.

Sectional annotation method can be used independently or combined with the ichnographic annotation method.

Figure 4.5.5 shows an example of ichnographic construction drawings of beams with sectional annotation.

3. Ichnographic construction drawing of slabs with beams

In a building, slabs are horizontally placed members separating the vertical space. The stress on slabs is similar to that on the beam and slabs are the typical flexural members. Reinforced concrete slabs are currently the most widely used slabs. According to the situation of force and force transmission, reinforced concrete slabs may be classified into slabs with and without beams. This section focuses on the ichnographic construction drawing of slabs with beams for the design of floor or roof slabs supported by beams.

1) Representation method of ichnographic construction drawings of slabs with beams

Ichnographic construction drawings of the slab with beams are to show slab information with the ichnographic annotation method in the plan layout of floor and roof slabs. Ichnographic annotations mainly contain general annotations of slab units and specific annotations of slabs at the support.

For the convenience of design and construction, the coordinate system in the structural layout plan is set as follows:

(1) When gridlines are orthogonally arranged, the positive direction of X is from left to right and the positive direction of Y is from bottom to top.

(2) If the grid is rotated at a certain position, the direction of the local coordinate system shall be rotated correspondingly.

(3) If grid lines are arranged radially, the tangential direction is the direction of the X-axis and the radial direction is the direction of the Y-axis.

Moreover, if the plan layout is too complicated, some key items, such as the intersection of folded gridlines, the core area of the radial layout, and the directions of the coordinate sys

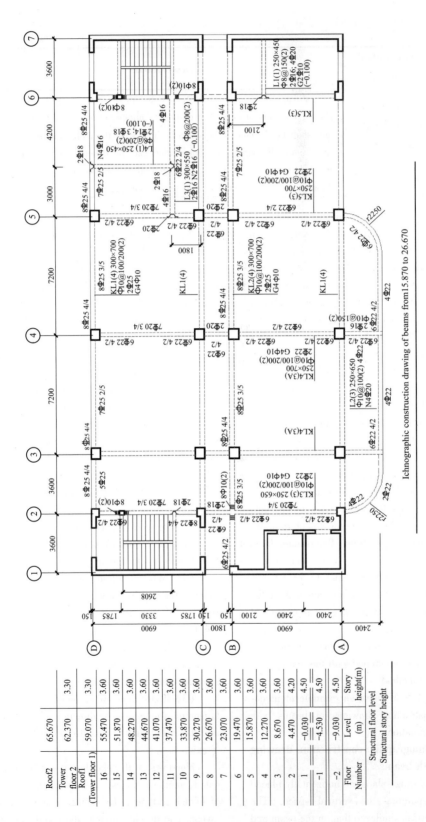

Figure 4.5.4　An example of ichnographic construction drawings of beams with ichnographic annotation

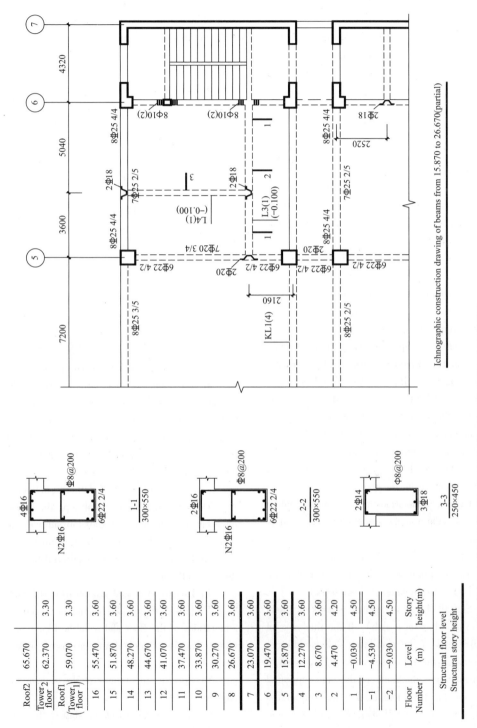

Figure 4.5.5 An example of ichnographic construction drawings of beams with sectional annotation

tem shall be specified by designers and denoted clearly in the drawings.

2) Slab number

Slab number is composed of type code and serial number, as shown in Table 4.5.5.

Slab number Table 4.5.5

Slab types	Type codes	Serial numbers
Floor slabs	LB	××
Roof slabs	WB	××
Cantilevered slabs	XB	××

3) General annotations of slab units

General annotations shall include: slab number, slab thickness, top continuous longitudinal bars, and bottom longitudinal bars, and the level difference shall be added when the slab top-level is different from that of the structural floor.

For ordinary floor slabs, one slab unit is a slab covering one span in two directions. For slabs with closely spaced ribs, one slab unit is a slab covering one span of primary beam (frame beam) in two directions. Ribs, which are not primary beams, will not be considered as primary beams. Slab units shall be numbered one by one. For the same slabs, the general annotation may be denoted on one slab unit and the same number (with a circle) may be denoted on the other slabs. If the level difference exists, it shall be denoted on the slab unit.

General annotations of slab units contain three compulsory items and one optional item as follows:

(1) Slab number (compulsory)

The slab number is shown in Table 4.5.5.

(2) Slab thickness (compulsory)

Slab thickness is denoted as $h = \times\times\times$ (normal to the plane of a slab). If the thickness of the free end of the cantilever slab is different from that of the supported end, "/" is used to distinguish the thickness of the slab-supported end from the thickness of the free end of the slab: $h = \times\times\times/\times\times\times$. If slab thickness has been summarized in notes, it can be omitted here.

(3) Longitudinal bars (compulsory)

Longitudinal bars are denoted separately as bottom longitudinal bars and top continuous longitudinal bars (when continuous bars are not arranged at the top, no note is required). "B" indicates bottom longitudinal bars; "T" indicates top continuous longitudinal bars; "B&T" indicates bottom and top bars; "X" indicates longitudinal bars in the direction of X-axis; "Y" indicates longitudinal bars in the direction of Y-axis; "X&Y" indicates that longitudinal bars in both directions are the same.

For one-way slabs, distribution bars can be summarized in drawing notes and no note in general annotations is required.

If constructional bars are designed in some slabs (such as at the bottom of cantilever slabs XB), "Xc" or "Yc" can be used to show bars in the direction of the X-axis and bars in the direction of the Y-axis, respectively.

If radial bars are placed in the direction of the Y-axis (the tangential direction is the direction of the X-axis and the radial direction is the direction of the Y-axis), designers shall denote the setting outlines for setting bar spacing.

If two diameters longitudinal bars are designed, they shall be staggered and denoted as ϕ xx/yy@ ×××, which means that the spacing of adjacent bars (xx and yy) is ××× and that the spacing of adjacent bars with the same diameter (xx or yy) is twice of ×××.

(4) Level difference of slab top (optional)

The level difference of slab top, the difference between the level of slab top surface and that of structural floor top surface, shall be denoted in brackets. No note is required if there is no level difference.

4) Example of general annotations of slab units

Table 4.5.6 provides examples of general annotations of the floor slab. Table 4.5.7 provides examples of general annotations of cantilever slabs.

5) Specific annotations of the slab at the support

Specific annotations of the slab at the support shall include the following items: top discontinuous longitudinal bars in the slab at the support and top main bars in the cantilever slab.

Examples of general annotations of floor slabs   Table 4.5.6

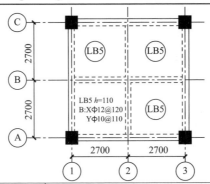

Methods	Denotation contents	Illustrations
General annotations	LB5	**Slab Number** The floor slab number is 5
	$h = 110$	**Slab thickness** The slab thickness of the No. 5 slab is 110
	B: X ϕ 12@120 Y ϕ 10@110	**Longitudinal bars** The bottom longitudinal bars are ϕ 12@120 in the direction of the X-axis and ϕ 10@110 in the direction of the Y-axis. No continuous longitudinal bars are arranged at the top

Examples of general annotations of cantilever slabs   Table 4.5.7

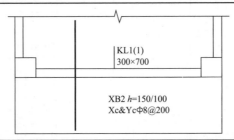

Methods	Denotation contents	Illustrations
General annotations	XB2	**Slab Number** The number of cantilever slabs is 2
	$h = 150/100$	**Slab thickness** The thickness at the supported end is 150 and the thickness at the free end is 100
	B: Xc&Yc ϕ 8@200	**Longitudinal bars** The bottom constructional bars are ϕ 8@200 in both directions

Specific annotations of bars in the slab at the support shall be denoted in the first span of all spans with the same configuration (if reinforcement bars in the cantilever slab are arranged separately, specific annotations of cantilever slab are required). In the first span of all spans with the same configuration (or cantilever beam), a medium line with a proper length (if continuous bars are arranged at the top of cantilever slab or short-span slab, this line shall extend to the far side of the support or cover the whole span of the short span) normal to slab support (beam or wall) is used to represent top discontinuous bars at the support, bar number (such as ① and ②), the amount of bars, the number of transverse spans covered (denoted in brackets, if the number of spans is one, no note is required). In addition, whether bars shall extend to the free end of the cantilever shall be denoted above the medium line. (××) indicates the number of transverse spans the main bar covered; (××A) indicates the number of transverse spans covered and one cantilever end covered as well; (××B) indicates the number of transverse spans covered and both cantilevers end covered as well.

The development length of top discontinuous bars at the support, measured from the centerline of the support, is denoted below the line.

If the development lengths on both sides of interior support are the same, only one side shall be noted below the line, as shown in Figure 4.5.6.

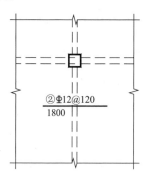

Figure 4.5.6 An example of top discontinuous bars in the slab at the support symmetrically arranged on both sides of the support

If the development lengths on both sides of interior support are different, they shall be noted separately on each side below the line, as shown in Figure 4.5.7.

As for the bars which extend to the far side of support to cover the whole span or to the free end of the cantilever slab, no note for the development length is required. Only the development length on the other side shall be denoted,

as shown in Figure 4.5.8.

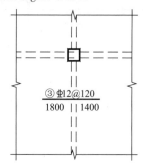

Figure 4.5.7 An example of top discontinuous bars in the slab at the support asymmetrically arranged on both sides of the support

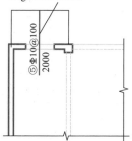

Figure 4.5.8 An example of top discontinuous bars in the slab at the support covering the whole span or extending to the free end of the cantilever slab

6) Example of ichnographic construction drawings of slabs with beams with ichnographic annotation.

An example of ichnographic construction drawings of slabs with beams with ichnographic annotation is shown in Figure 4.5.9.

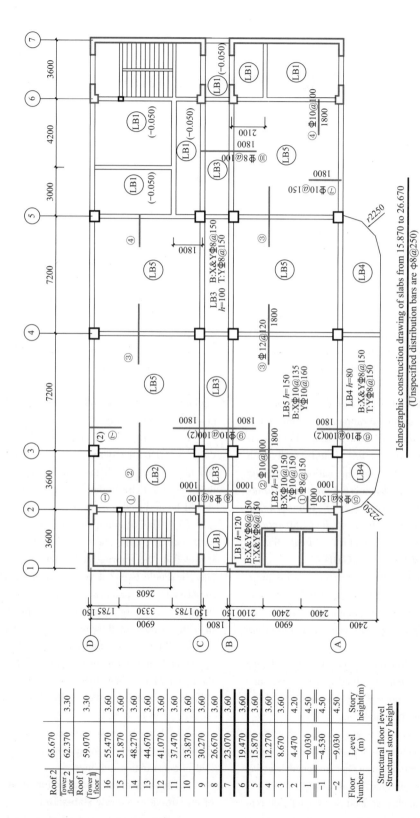

Figure 4.5.9 An example of ichnographic construction drawings of slabs with beams

4.6 Ichnographic Construction Drawings of Stairs

Stairs are used as a member of vertical traffic between floors in a building. Stairs may be classified into straight flight stairs, double flight stairs, three flight stairs, four flight stairs, double-separating stairs, cross-zigzag stairs, etc. According to stair structures, stairs may be classified into slab stairs, girder stairs, cantilevered stairs, etc. The section mainly introduces ichnographic construction drawings of cast-in-situ concrete slab-stairs.

1. Overview

Construction drawings designed according to IRM (ichnographic representation method) usually contain two main parts: ichnographic construction drawings of stairs and standard detail drawings.

IRM of stair slabs includes ichnographic annotation, sectional annotation, and tabulate1 annotation. IRM of landing slabs, stair beams, and stair columns should comply with the drawing collection for national building standard design, *Drawing Rules and Standard Detailing Drawings of Ichnographic Construction Drawings of R. C. Structures (Cast-in-situ R. C. Frames, Shear Walls, Beams, and Slabs)* (**16G101-1**).

To complete the construction process properly according to drawings, the information closely related to ichnographic construction has to be specified in specific construction drawings as below:

(1) The reference number of the adopted ichnographic standard drawing collection shall be specified to prevent any misuse of the drawing collection after updating.

(2) The concrete grade and steel grade of reinforcement bars in stairs shall be denoted to identify the minimum anchorage length and lap length of longitudinal tension bars. When the mechanical anchorage mode is adopted, designers shall specify the type of anchorage, minimum dimensions of members, and quality requirements for mechanical anchorage.

(3) The environment type of stairs shall be denoted.

(4) If the type of stairs is ATa, ATb, ATc, CTa, or CTb, the seismic grade should be given by designers according to situations of specific projects.

(5) If more than one option is provided in the standard detailed drawings, the location and the corresponding detailed practice of each option shall be clarified. Designers should denote the structures selected for each detailing practice in ichnographic construction drawings. If the same detailing practice is adopted in the majority of structures, it may be specified in drawing annotation, whereas other options should be denoted in the drawing.

(6) If the type of stairs is ATa, ATb, CTa, or CTb, the detailing of slide bearings should be specified. If the detailing practice is different from those in the drawing collection, it should be designed separately.

(7) The detailing drawing of embedded parts connecting the stairs and handrails is not included in the drawing collection. In the design, it should be noted that the detailing drawing of embedded parts connecting the stairs and handrails are shown in the architectural design drawing or relevant drawing collections for national building standard design.

(8) In a given project, if a modification of the standard detailing drawings in the drawing collection is required, the details of the modification should be denoted.

(9) If there are special requirements in a given project, the requirements should be specified in construction drawings.

2. Ichnographic construction drawings of cast-in-situ concrete slab-stairs

1) Representation method of ichnographic

construction drawings of cast-in-situ concrete slab-stairs

Representation methods of ichnographic construction drawings of cast-in-situ concrete slab-stairs include ichnographic annotation, sectional annotation, and tabulated annotation.

This section introduces the representation method of stair slabs. Related details may be acquired in the drawing collection for national building standard design 16G101-1 for the representation method of landing slabs, stair beams, and columns related to stairs.

The plan layout of stairs should be drawn in a drawing according to an appropriate scale. If necessary, sectional drawings should be provided as well. For the convenience of construction, structural floor levels, structural story heights, and corresponding story numbers should be denoted in ichnographic construction drawings of slab-stairs.

2) Types of stairs

There are 12 types of stairs, as shown in Table 4.6.1.

Types of stairs　　　　　　　　　　　　　　Table 4.6.1

Type codes of stair slabs	Scopes of application		To be or not to be considered in the global seismic analysis
	Seismic detailing	Structures	
AT	None	Shear wall & masonry structures	No
BT			
CT	None	Shear wall & masonry structures	No
DT			
ET	None	Shear wall & masonry structures	No
FT			
GT	None	Shear wall & masonry structures	No
ATa	Applicable	Frame structures & the frame part in frame-shear wall structures	No
ATb			No
ATc			Yes
CTa	Applicable	Frame structures & the frame part in frame-shear wall structures	No
CTb			No

Note: The slide bearing sitting on the stair beam should be at the lower end of ATa and CTa, whereas the slide bearing sitting on the cantilever slab should be at the lower end of ATb and CTb.

(1) Stair number

The numbering of stairs is composed of the type code and serial numbers, such as AT××, BT××, and ATa××.

(2) AT ~ ET slab-stairs (Figure 4.6.1) has the following features:

① Type codes AT ~ ET indicate the stair slab with upper and lower supports. Except for the flight which is the main part of a stair slab, the stair slab may be composed of a lower landing slab, upper landing slab, and middle landing slab.

② Section shapes of AT ~ ET stair slabs are described as follows:

AT stair slab consists of a flight only;

BT stair slab consists of a lower landing slab and a flight;

CT stair slab consists of a flight and an upper landing slab;

DT stair slab consists of a lower landing

slab, a flight, and an upper landing slab;

ET stair slab consists of a lower flight, a middle landing slab, and an upper flight.

③ Two ends of AT ~ ET stair slabs are supported by (lower and upper) stair beams.

④ Type codes, slab thickness, top and bottom longitudinal, and distribution bars in AT ~ ET stairs are denoted by designers in ichnographic construction drawings. Horizontal projection length of top longitudinal bars in stair slabs developed to the span may be acquired in the corresponding standard detailing drawings. It is not required in design drawings, but it should be checked by designers. If horizontal projection length in the standard detailing drawings can not meet requirements in specific projects, it should be denoted by designers separately.

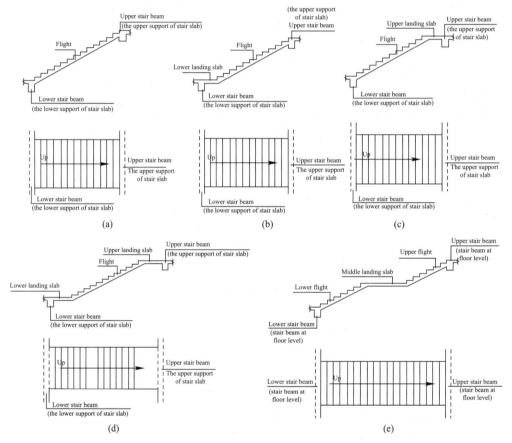

Figure 4.6.1 Cross-section and supports of AT ~ ET stairs
(a) Type AT;(b) Type BT;(c) Type CT;(d) Type DT;(e) Type ET

(3) Features of FT and GT slab-stairs

① Type codes FT and GT (Figure 4.6.2) denote two flights with intermediate landing slabs and landing slabs at the floor level.

② The components of FT and GT stair slabs are described as:

FT consists of an intermediate landing slab, two flights, and two landing slabs at floor levels.

GT consists of an intermediate landing slab and two flights.

③ Support patterns of FT and GT stair slabs (Table 4.6.2) are described as:

FT: The intermediate landing slab is supported on three edges and the landing slab at floor level is also supported on three edges.

GT: The intermediate landing slab is supported on three edges and the other end of the flight is supported by the stair beam.

Support patterns of FT and GT stair slabs Table 4. 6. 2

Types of stair slabs	Intermediate landing slabs	End of flight (at floor level)	Landing slab at the floor level
FT	Supported on three edges	—	Supported on three edges
GT	Supported on three edges	Supported by stair beam	—

④ Type codes, slab thickness, top, and bottom longitudinal bars, and distribution bars in FT and GT stair slabs are denoted by designers in ichnographic construction drawings. Top transverse bars in landing slabs of FT and GT stairs and the development length are denoted with specific annotation in plan drawings. Horizontal projection length of top longitudinal bars in stair slabs developed to the span can be consulted in the corresponding standard detailing drawings. It is not required in design drawings, but it should be checked by designers. If horizontal projection length in the standard detailing drawings can not meet the requirements in specific projects, it should be denoted by designers separately.

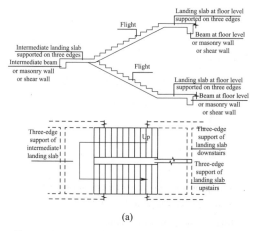

Figure 4.6.2 Cross-section and supports of FT and GT stairs
(a) Type FT;(b) Type GT

(4) Features of ATa and ATb slab-stairs

① ATa and ATb stairs (Figure 4.6.3) are stairs with slide bearings. The stair slab consists of only a flight and the upper end is supported by the stair beam. The lower end of ATa stair slab is supported by the slide bearing sitting on the stair beam. The lower end of ATb stair slab is supported by the slide bearing sitting on the cantilever slab.

② Designers should determine the materials. The backing plate of the slide bearing may be a PTFE or steel plate or a plastic plate with a minimum thickness of no less than 0.5. Other materials that can ensure useful sliding may be selected. Connection details of backing plates should be determined by designers.

③ Reinforcement in two directions should be placed at both the top and bottom of ATa and ATb stair slabs.

(5) Features of ATc slab-stairs

① The stair slab consists of only a flight and is supported by stair beams at both ends.

② The landing slab can be connected or disconnected from the main structure.

③ The thickness of the stair slab should be determined by calculation and is recommended to be no less than 140mm. Reinforcement bars should be placed both at the top and at the bottom of the stair slab.

④ Boundary elements (embedded beams) are placed on both sides of the stair slab and their width is 1.5 times of slab thickness. When a seismic grade is Grade Ⅰ or Ⅱ, at least 6 longitudinal bars shall be placed in the boundary element. When the seismic grade is Grade Ⅲ or Ⅳ, at least 4 longitudinal bars should be placed in boundary elements. The diameter of longitudinal bars should not be less than ϕ12 and the diameter of longitudinal main bars in the stair slab. The diameter of stirrups should not be less than ϕ6 with the spacing no more than 200.

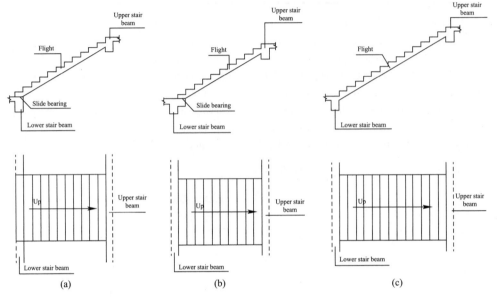

Figure 4.6.3 Cross-section and supports of ATa ~ ATc stairs
(a) Type ATa;(b) Type ATb;(c) Type ATc

Reinforcement bars in two directions should be placed at the top and bottom of the landing slab.

⑤ If ATc stairs act as diagonal bracing members, reinforcement bars should be hot-rolled bars that meet seismic performance requirements. The ratio of measured tensile strength to measured yield strength should be no less than 125 and the ratio of the measured yield strength to the characteristic yield strength should be no higher than 1.3. Measured elongation under ultimate tension force should be no less than 9%.

(6) Features of CTa and CTb slab-stairs

① CTa and CTb stairs are stairs with slide bearings. The stair slab consists of a flight and an upper landing slab, and the upper end is supported by the stair beam. The lower end of CTa stair slab is supported by the slide bearing sitting on the stair beam. The lower end of CTb stair slab is supported by the slide bearing sitting on the cantilever slab.

② Designers should determine the materials. The backing plate of the slide bearing may be a PTFE or steel plate or a plastic plate with a minimum thickness of no less than 0.5. Other materials that can ensure useful sliding may be chosen as well. Connection details of backing plates should be determined by designers.

③ Reinforcement bars in two directions should be placed at the top and bottom of CTa and CTb stair slabs.

(7) Detailing of stair beams

If stair beams are supported by stair columns, detailing should comply with the detailing practices of frame beam KL in 16G101-1, and the stirrup densification zone is recommended to cover the whole length of the beam.

(8) Adjustment of the height of the rises

In architectural design, the thickness of the finishing layer of the ground floor, landing slab at floor level, and intermediate landing slab are usually different from those of stairs. To keep every rise the same after the finishing layer is finished, the rises of the bottom and top risers of different kinds of stairs should be increased or decreased.

3) Ichnographic annotation method

The ichnographic annotation method, including the general annotation and peripheral annotation, is to denote sectional dimensions and reinforcement details in the plan layout of stairs in construction drawings.

(1) General annotation

General annotation of stairs includes the following five items.

① Type code and a serial number of stair slabs, e. g. AT××.

② The thickness of stair slabs shall be denoted as $h = ×××$. If a stair slab consists of both flight and landing slabs with different thicknesses, the thickness of flight should be denoted firstly and the thickness of the landing slab may be denoted in brackets with the prefix "P".

e. g. $h = 130$ (P150) indicates that the flight thickness is 130 and that the landing slab thickness is 150.

③ The total height of flights and the number of risers shall be separated by "/".

④ Top and bottom longitudinal bars in supports of stair slabs shall be separated by ";".

⑤ Distribution bars in stair slabs. The specific value of distribution bars is denoted with the prefix "F". It may be shown in drawing annotation as well.

e. g. An example of a complete annotation of type and reinforcement details of stairs (Type AT) in a plan layout is shown in Table 4. 6. 3.

An example of a complete annotation of type and reinforcement details of stair slabs in a plan layout

Table 4. 6. 3

Serial numbers	Items	Illustrations
1	AT1	Type code and serial number of the stair slab
2	$h = 120$	Stair slab thickness
3	1800/12	The total height of the flight/the number of risers
4	$\Phi 10@200; \Phi 12(@150$	Top longitudinal bars; bottom longitudinal bars
5	F$\Phi 8@250$	Distribution bars in stair slab

⑥ For ATc stairs, longitudinal bars and stirrups in boundary elements on both sides of the stair slab should be denoted.

(2) Peripheral annotation

Peripheral annotation of stairs includes plane dimensions of staircases, structural floor levels, intermediate structure levels, directions of travel plane dimensions of stair slabs, reinforcement in landing slabs, reinforcement bars in stair beams, and stair columns.

4) Sectional annotation method

Sectional annotation method, including the ichnographic annotation and sectional annotation, is to draw the plan layout and section drawing of stairs in ichnographic construction drawings of stairs.

The items denoted in the plan layout of stairs include plane dimensions of staircases, structural floor levels, intermediate structure levels, directions of travel, plane dimensions of stair slabs, type codes and serial numbers of stair slabs, reinforcement in landing slabs, reinforcement in stair beams and stair columns.

The items denoted in section drawings of stairs include general annotation of stair slabs, the numbering of stair beams and stair columns, horizontal and vertical dimensions of stair slabs, structural floor levels, and intermediate structural levels.

General annotation of stair slabs includes the following four items.

(1) Type code and a serial number of stair slabs, e.g. AT××.

(2) The thickness of stair slabs is denoted as $h=×××$. If a stair slab consists of both flight and landing slabs with different thicknesses, the thickness of flight should be denoted first and the thickness of the landing slab may be denoted in brackets with the prefix "P".

(3) Reinforcement in stair slabs. Top and bottom longitudinal bars in stair slabs should be denoted and separated by ";".

(4) Distribution bars in stair slabs. The specific value of distribution bars is denoted with the prefix "F". It may be added in drawing annotation as well.

e.g. An example of a complete annotation of reinforcement details of stair slabs in sectional drawings is shown in Table 4.6.4.

An example of a complete annotation of reinforcement details of stair slabs in section drawings   Table 4.6.4

Serial numbers	Items	Illustrations
1	AT1	Type code and serial number of the stair slab
2	$h=120$	Stair slab thickness
3	$\phi 10@200; \phi 12(@150$	Top longitudinal bars; bottom longitudinal bars
4	F$\phi 8@250$	Distribution bars in the stair slab

(5) For ATc stairs, longitudinal bars and stirrups in boundary elements on both sides of the stair slab should be denoted.

5) Tabulated annotation method

The tabulated annotation method is to denote sectional dimensions and reinforcement details of stairs in a table in construction drawings.

The requirements of the tabulated annotation method are the same as those of the sectional annotation method except that the reinforcement details of stair slabs shall be denoted in a table.

Tabulated annotation of stair slabs is shown in Table 4.6.5.

Dimensions and reinforcement details of stair slabs   Table 4.6.5

Stair slab numbers	Total height of flight/number of risers	Slab thickness h	Top longitudinal bars	Bottom longitudinal bars	Distribution bars
AT1	1800/12	120	$\phi 10@200$	$\phi 12(@150$	$\phi 8@250$

Note: For ATc stairs, longitudinal bars and stirrups in boundary elements on both sides of the stair slab should be denoted.

6) Ichnographic annotation method and application of AT stairs

(1) If the stair slab between two stair beams only consists of flight, the stairs can be considered as AT stairs, in which stair beams serve as the supports of flight at both ends. If stairs meet these conditions, they can be considered as Type AT.

(2) Ichnographic annotation method of AT stairs is shown in Figure 4.6.4. General annotation of AT stairs includes the following five items:

① AT××, type code and serial number of the stair slab;

② Thickness of stair slab, h;

③ Height of flight, H, and the number of risers ($m+1$);

④ Top and bottom longitudinal bars in stair slab;

⑤ Distribution bars in the stair slab. A design example is shown in Figure 4.6.5.

(3) Distribution bars in the stair slab could be directly denoted in the drawing or specified in annotations

(4) The reinforcement annotation of landing slabs PTB, stair beams TL and stair columns TZ may be acquired from *Drawing Rules and Standard Detailing Drawings of Ichnographic Representing Method for Construction Drawings of R. C. Structures* (Cast-in-

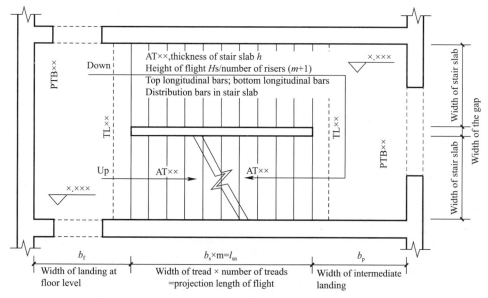

Figure 4.6.4 Plan layout of stairs from level x. xxx to level x. xxx

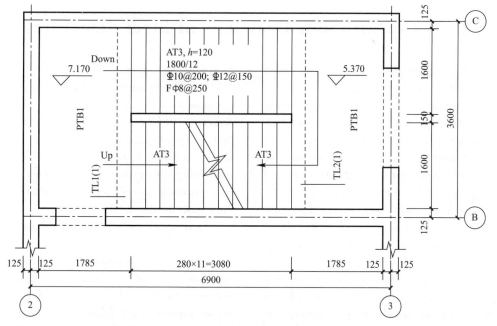

Figure 4.6.5 An example of the ichnographic annotation method of AT stairs

situ R. C. Frames, Shear Walls, Beams, and Slabs)(**16G101-1**).

5) Reinforcement details of AT stair slabs are shown in Figure 4.6.6.

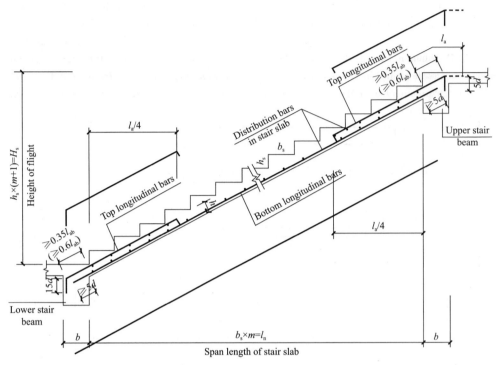

Figure 4.6.6　Reinforcement detailing of AT stair slabs

4.7　Construction Drawings of Steel Structures

A steel structure has high strength, small weight, high security, and convenient installation and manufacture and is widely applied in many buildings, including bridges, truss, industrial plants, and high-rise buildings.

A steel structure is the bearing structure composed of different kinds of formed steels and steel plates connected in various ways, including welding, bolts, and rivets. Thus, the items in a steel structure drawing generally include steel member type, shape, size, and connection type.

Similar to reinforced concrete structure drawings, steel structure drawings include different kinds of symbols, drawing numbers, illustrations, and national standards.

This chapter mainly introduces the relevant requirements of steel structures in ***Building Structure Drafting Standard*** (GB/T 50105—2010), including denotation methods of steel plates, section steel, bolts, holes, rivet welds, and welding connection ways. In addition, the requirements of common steel structures and relevant cases are discussed.

1. Denotation of steel plates and section steel

According to the national and industrial standards, the denotation methods of steel plates and section steels in steel structures shall be under the requirements in Table 4.7.1.

Denotation Method of Section Steel Table 4.7.1

No.	Names	Sections	Denotations	Illustrations
1	Steel plate	▬	$\dfrac{-b \times t}{l}$	$\dfrac{-\text{Width} \times \text{thickness}}{\text{length}}$
2	Equal leg angle steel	 L	∟ $b \times d$	"b" is the width; "d" is thickness
3	Unequal angle bar	 L	∟ $B \times b \times d$	"B" is the width of a long limb; "b" is the width of a short limb; "t" is limb
4	I-section steel	I	$Q\,\mathrm{I}\,N \quad \mathrm{I}\,N$	I-section steel is denoted with "Q"
5	Channel steel	[	$[\,N \quad Q\,[\,N$	
6	Square steel	 ▨	□ b	
7	Flat steel	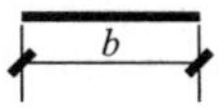 ⊢ b ⊣	▬ $b \times h$	
8	Round steel	⊘	$\varnothing d$	
9	Steel tube	 ○	$DN\text{xx}$ $d \times t$	
10	Thin-wall square steel tube	□	$B\,\square\,b \times t$	
11	Thin-wall angle steel with equal legs	 L	$B\,\llcorner\,b \times t$	
12	Thin-wall flanging angle steel with the same kind of limb	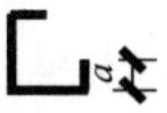 ⌊ a	$B\,\llcorner\,b \times a \times t$	
13	Thin-wall channel steel	 [h	$B\,[\,h \times b \times t$	
14	Thin-wall flanging channel steel	 ⌊ a	$B\,[\,h \times b \times a \times t$	
15	Thin-wall flanging Z-section steel	 ⌐⌙	$B\,\lrcorner\,h \times b \times a \times t$	

No.	Names	Sections	Denotations	Illustrations
16	T-section steel	T	TW×× TM×× TN××	TW indicates wide-flange T-section steel; TM indicates medium-flange T-section steel; TN indicates narrow-flange T-section steel
17	H-section steel	H	HW×× HM×× HN××	HW indicates wide-flange H-section steel; HM indicates medium-flange H-section steel
18	Crane steel rail		 QU××	Product specifications and models
19	Light rail and steel rail		××kg/m	

2. Denotation of bolts, holes, and electric welding rivets

The steel structure connection is generally performed by bolting and welding. Bolted connection is mainly used in the installation of steel structures and the connection of different parts.

Bolted connection often adopts common bolts and high-tensile bolts. Common bolts are generally made of Q235 steel and high-strength bolts are made of high-strength steel obtained after heat treatment.

Common bolts are classified into Grades A, B, and C. As for the bolts of Grade A and Grade B, the bore diameters are 0.3 – 0.5mm larger than bolt core diameter d. As for bolts of Grade C, its diameter is 1.5 – 3mm larger than bolt core diameter d.

High-strength bolts may be classified into friction type and bearing type according to their mechanism. The bore diameter of friction-type bolts is 1.5 – 2.0mm larger than screw diameter d; whereas the bore diameter of bearing-type bolts is 1.0 – 1.5mm larger than screw diameter d.

3. Denotation of welds

In a steel structure, welding is the main connection method since it does not change the member section. Welding is simple and convenient. The connection type of section steel is different from that of steel plates, so their welding patterns are also different. Welding connection can be classified into butting connection, lap connection, and T-type connection according to the relative positions of connected members. In addition, the welds are classified into butt weld and fillet weld, as shown in Figure 4.7.2.

Denotation of bolts, holes, and electric welding rivets   Table 4.7.2

No.	Names	Drawings	Descriptions
1	Common/normal bolt	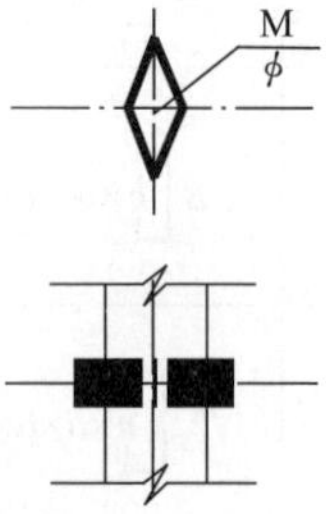	"+" indicates a positioning line; "M" indicates bolt model; "ϕ" indicates bolt diameter; "d" indicates diameter

Continued

No.	Names	Drawings	Descriptions
2	High-strength bolt	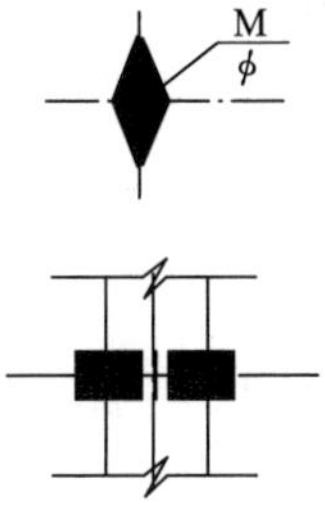	
3	Mounting bolt	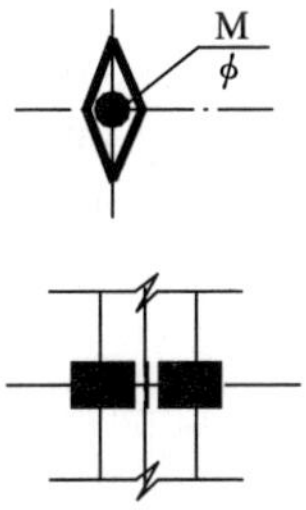	
4	Cinch bolt	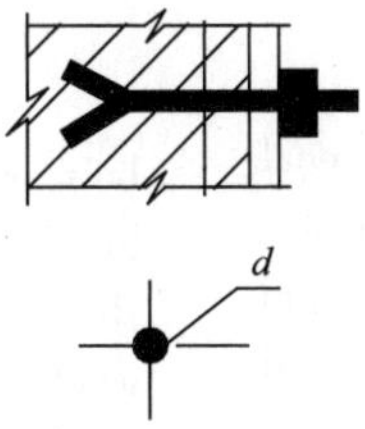	"+" indicates a positioning line; "M" indicates bolt model; "ϕ" indicates bolt diameter; "d" indicates diameter
5	Round bolt hole	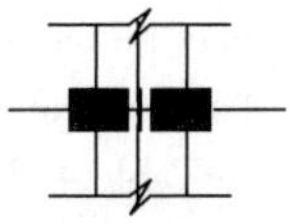 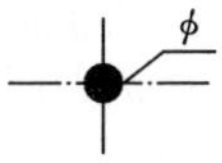	
6	Long round bolt hole	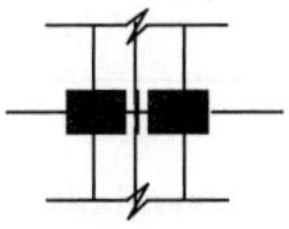 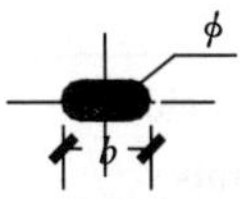	
7	Electric welding rivet	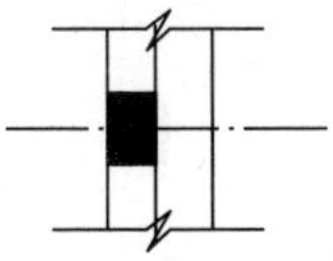 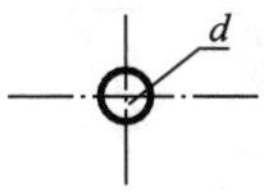	

Fillet welds can be classified into joint fillet welds and interlude fillet welds. Welding patterns include flat welding, horizontal welding, vertical welding, and overhead welding according to the spatial position of welds.

In a steel structure drawing, the welding type, welding seam form, position, size, and welding method shall be denoted with weld symbols.

1) Welding symbols

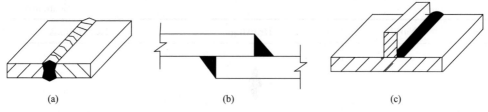

Figure 4.7.1 Welding connection type
(a) Butt connection;(b) Lap connection;(c) T type connection

According to the ***Representation Method of Welding Symbols*** (GB/T 324—2008), weld symbols are generally composed of basic symbols and outgoing lines. If necessary, auxiliary symbols, supplementary symbols, and weld size may be added.

Indexing line is generally composed of an arrow and two basic lines, including one fine solid line and one fine dotted line, as shown in Figure 4.7.2. The dotted line can be drawn below or above the solid line. Welding symbols are labeled above or below the basic lines. The whole arrow indicates the position of the weld.

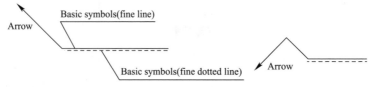

Figure 4.7.2 Indexing lines

Basic symbols indicate the cross-sectional shape of welds. For example, "V" indicates the butt welding seam of the V-groove, and "⌓" indicates fillet weld. Basic symbols and labels for welds are illustrated in Table 4.7.3.

Auxiliary symbols indicate the shape features of a welding surface and will be labeled on the corresponding position along with basic symbols. For example, "$\overline{V}$" indicates that the over-high part of the V-shaped welding seam surface should be processed to be on a level with the weld surface. If it is not necessary to explain the welding seam surface, auxiliary symbols may be omitted. Commonly used auxiliary symbols and labels for welding seams are illustrated in Table 4.7.4.

Commonly used basic symbols and denotation examples of welding seams

Table 4.7.3

Names of welding seams	Types of welding seams	Basic symbols	Examples
I-type welding seam		‖	
Single-side V-shaped welding seam		V	

Continued

Names of welding seams	Types of welding seams	Basic symbols	Examples
Bilateral V-shaped welding seam	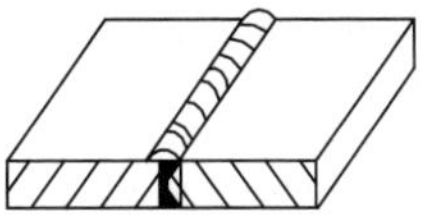	V	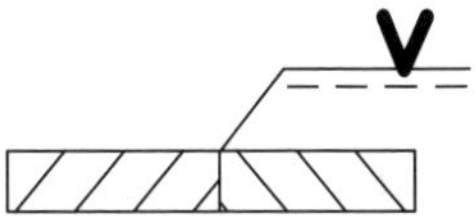
Bilateral and single-side V-shaped welding seam	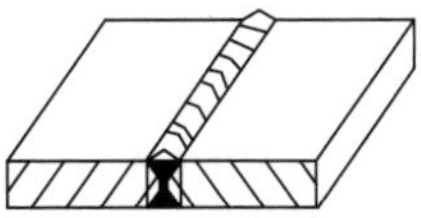	 K	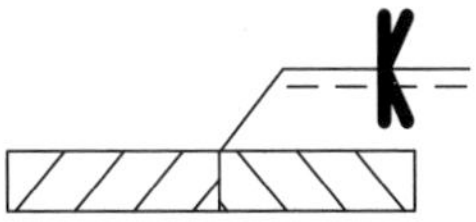
Double V-shaped welding seam	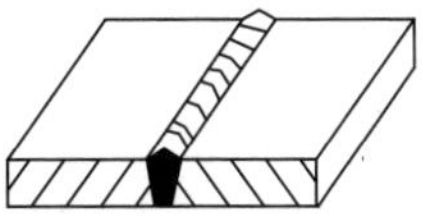	 X	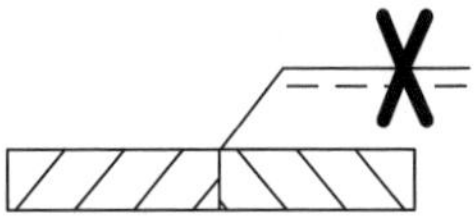
Fillet weld	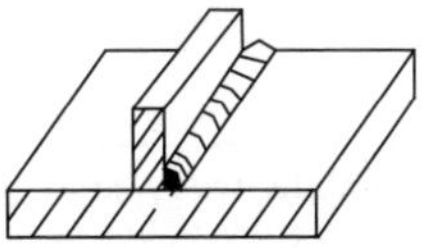	 △	

Auxiliary symbols and denotation examples of welds　　　Table 4.7.4

Auxiliary symbols	Denotation Examples	Descriptions
Flat symbol "—"	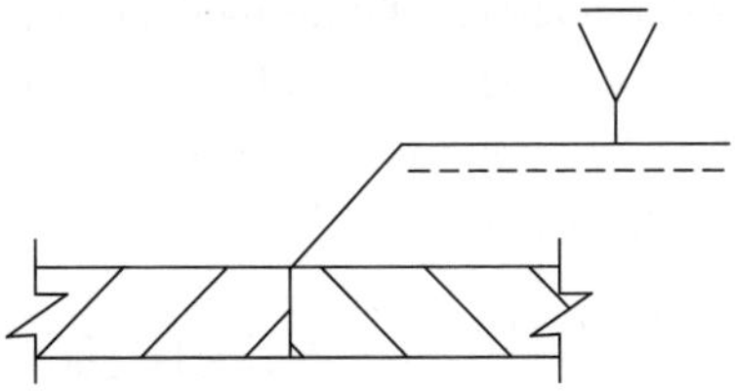	The bilateral V-shaped butt welding seam surface is smooth
Concave symbol "⌣"	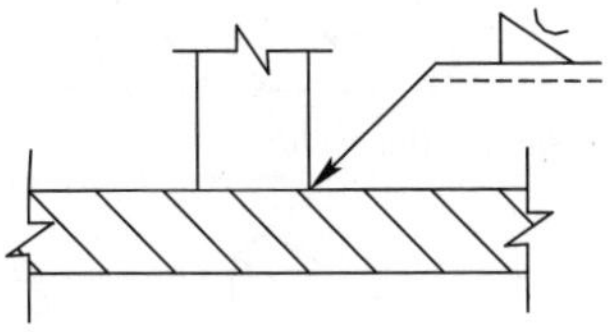	T-type connection fillet weld surface is caved
Convexity symbol "⌢"	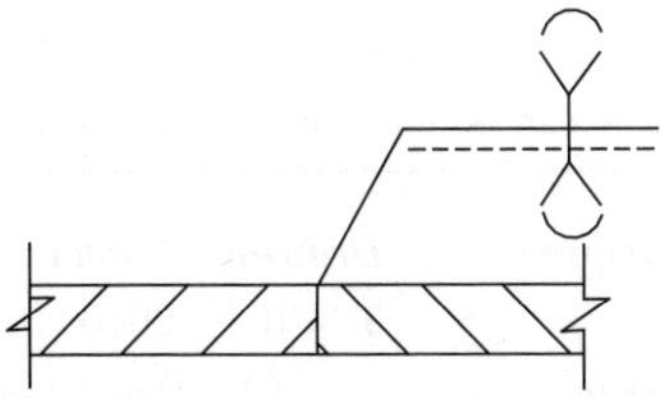	Double V-shaped butt welding seam heaves on the surface

Supplementary symbols are the symbols set to additionally indicate certain features of the weld and may be denoted at the corresponding positions along with the basic symbols. For example, "○" indicates peripheral weld. Table 4.7.5 is the illustration of commonly used supplementary symbols and labels for welding seam.

Commonly used supplementary symbols and denotation examples of welding seams

Table 4.7.5

Supplementary symbols	Denotation examples and descriptions
Pallet 	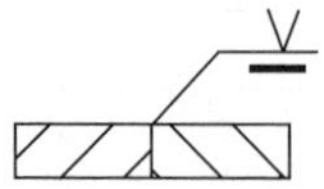 Bilateral V-shaped butt welding seam with bottom pallet
Trilateral welding seam ⊏	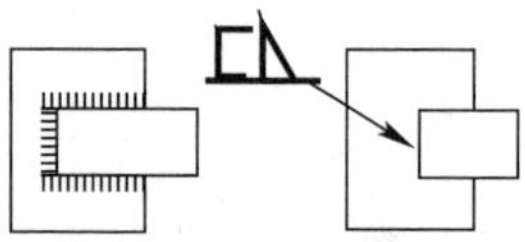 Fillet weld is carried out along the three sides of the weldment
Site welding	The symbol of site welding is a triangle flag, which is drawn at the turnover section of the outgoing line
Identical welding seam 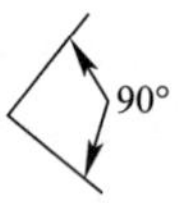 90°	In the same drawing, when the weld type, cross-section dimension, and auxiliary requirements of various welds are the same, it is necessary to select only one place to give the symbol and size for the weld and the symbol of identical welding seam shall be added. The symbol of an identical welding seam is 3/4 of arc, which is drawn at the turnover section of the outgoing line. In the same drawing, if there are various types of welds, welds may be given with the serial number according to their category. Among the same type of welds, it is necessary to select a position to give the symbol and size of welds. Capital letters (A, B, and so on) are used as the category serial number. 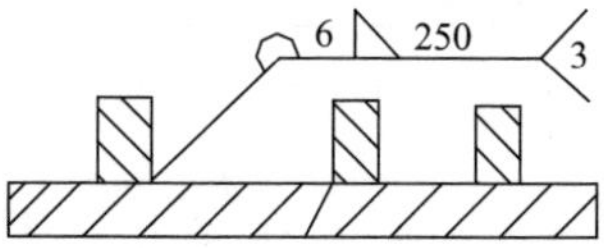 There are three same fillet welds
Around weld symbol ○	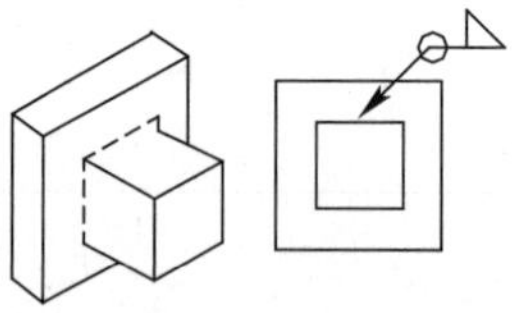 Fillet welding is carried out along the weldment periphery

2) Relevant specifications and requirements for weld denotation

The denotation for a butt-welded seam in *Drawing Standard of Building Structure* (GB/T 50105—2010) is provided as follows:

(1) Denotation of single-side weld

When an arrow is directed to the side of a single-side weld, the graphic symbol and size of the weld shall be denoted above the transverse line. When the arrow is directed to the other side of a single-side weld, the graphic symbol and size of the weld shall be denoted below the transverse line, as shown in Figure 4.7.3.

A peripheral weld is denoted with a circle drawn at the turning section of the outgoing line and the size (K) of a fillet weld shall be denoted.

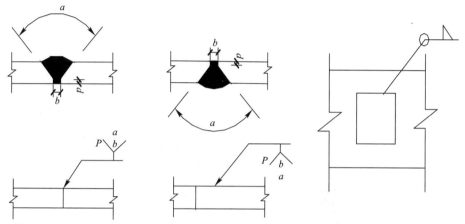

Figure 4.7.3 Denotation method of single-side weld

(2) The symbols and sizes of a double-side weld shall be denoted above and below the transverse line. The symbol and size on the arrow side are denoted above the transverse line, whereas the symbol and size on the other side are denoted below the transverse line. When the weld sizes on both sides are the same, it is only necessary to denote the symbol and size above the transverse line (Figure 4.7.4).

(3) The weld connecting three or more three weldments should not be denoted as a double-side welding seam. The symbol and size of the weld shall be respectively denoted (Figure 4.7.5).

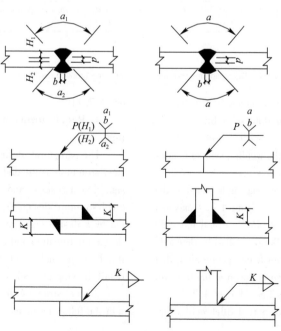

Figure 4.7.4 Denotation methods of double-side welds

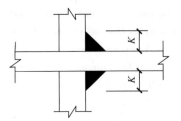
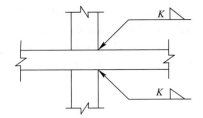

Figure 4.7.5 Weld denotation method

(4) When only one of the two weldments has a groove (such as a single-side V-shaped welding seam), the arrow must be directed to the weldment with the groove (Figure 4.7.6).

As for a single-side weld with a bilateral unsymmetrical groove, the arrow of the outgoing line must be directed to the weldment with the larger groove (Figure 4.7.7).

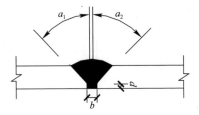
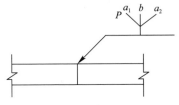

Figure 4.7.6 Denotation method of welds with an unsymmetrical groove

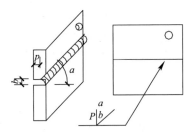
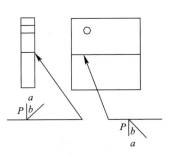

Figure 4.7.7 Denotation method of the weld with one grooved weldment

(5) When a weld is unequally distributed, it is better to add a middle solid line (visible weld) or refine line (invisible weld) at the welding section when the welding symbol is added (Figure 4.7.8).

(6) The symbol of long welds (such as the flange welding seam) in Figure 4.7.9 may not be denoted with the outgoing line and the weld size (K) around the fillet weld shall be denoted.

(7) The symbol for fusion penetration fillet welds is a circle, which is drawn at the turnover section of the outgoing line (Figure 4.7.10).

(8) Denotation of partial fillet weld (Figure 4.7.11).

4. Requirements of steel structure drawings

The structural design drafting process of steel structures is divided into steel structure design drawing stage and detail drawing stage. Steel structure design drawing shall be completed by a design organization with corresponding design qualification and the content and depth of the drawings shall meet the requirements of the detail drawing of steel structures.

A steel structure design drawing shall contain the following items:

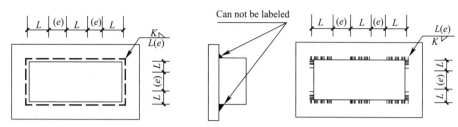

Figure 4.7.8　Denotation method of irregular welds

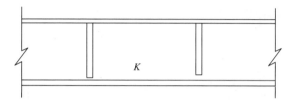

Figure 4.7.9　Denotation method of longer welds

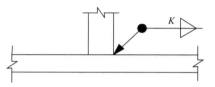

Figure 4.7.10　Denotation method of fusion penetration fillet welds

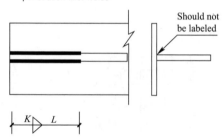

Figure 4.7.11　Denotation method of partial fillet welds

1) Design introduction shall cover the design basis, load, project type, project overview, type of steels, quality grade (if necessary, physical, mechanical property and chemical composition shall be added), specifications, weld quality grade, corrosion prevention, and fire precaution measurement.

2) Foundation plans and detail drawings shall contain the connection structure detail drawing between steel columns and concrete components.

3) Structural plan (including each floor and roof) layout drawing shall contain location, elevation, member position, serial number, index number, and so on. If necessary, purlin and wall beam layout drawing and key sectional drawing shall be drafted. As for spatial wireframes, the drawings of upper and lower chord and key cross-section shall be plotted.

4) Detail drawings of members and nodes:

(1) The simple steel beam and pillars can be denoted with the uniform detail drawing and tabulating method to indicate member steel type, size, specification, connecting nodes, and construction and installation requirements.

(2) Detail drawing of connecting nodes and other structural joints.

5. An example of steel structure drawing

Steel structure construction drawings are enlarged drawings of all the parts for structural elements. The parts shall be numbered according to the type, size, material properties, process requirements, weld quality grade, and so on. In addition, the transportation and installation capacity shall be considered.

Figure 4.7.12 shows the construction drawing of a steel structure awning (sourced from www.zhulong.com).

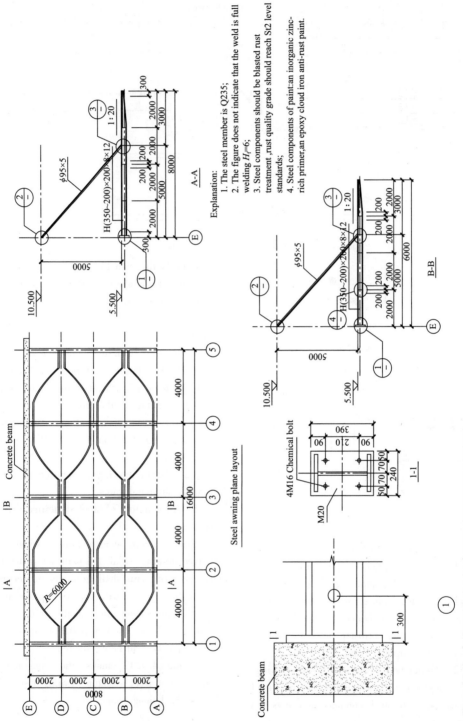

Figure 4.7.12 Construction drawing for a steel structure awning(one)

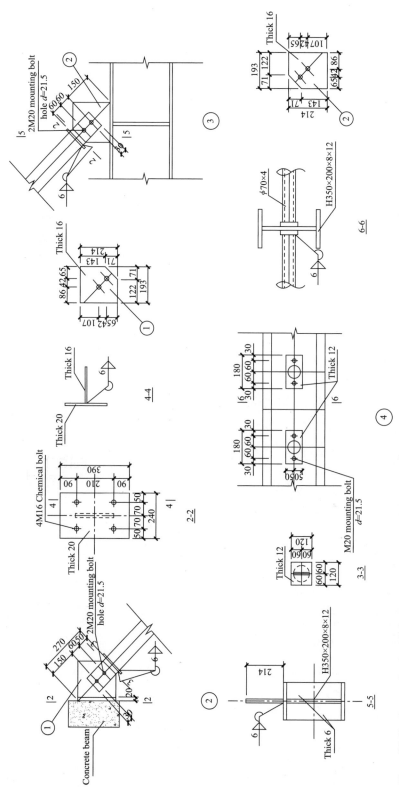

Figure 4.7.12 Construction drawing for a steel structure awning(two)

Chapter 5
Road Route Engineering Drawings

A road is a belt-like infrastructure for vehicles and pedestrians. Its basic components include roadbeds, pavements, bridges, culverts, tunnels, protection works, drainage facilities, etc. According to the composition and functional characteristics, roads can be classified into two kinds: highways and urban roads. The road located in the outskirts of a city and outside of a city is called a highway, whereas the one located within a city is called an urban road, as shown in Figure 5.0.1.

Figure 5.0.1 Road maps
(a)Expressway;(b)Arterial road;(c)Winding mountain road;(d)Urban expressway;(e)Urban thoroughfare;(f)Urban overpass

Road engineering is characterized by complex components and significant three-dimensional differences between length, width, and height and is largely affected by topography. In addition, it involves multiple disciplines. Unlike general engineering drawings, road engineering drawings are based on topographic maps (the plan drawings), longitudinal expansion section drawings (elevation drawings), and cross-sectional drawings (side views), most of which are arranged on individual sheets. The final design results of a road, including the spatial position, line types, and dimensions, are expressed based on plan drawings, longitudinal section drawings, and section drawings. This chapter introduces the denotation method, drawing features, and contents of road engineering. Road engineering drawings shall meet relevant provisions in ***Drawing Standards of Road Engineering*** (GB 50162—92).

5.1 Highway Route Drawings

A road is a narrow and long linear engineering structure built on the ground for vehicles and pedestrians. A road route refers to the centerline of the carriageway along the length of the road. The location and shape of a road are closely related to the terrain, topography, features, and geology in the construction site. A road route often has vertical height changes (uphill, downhill, and vertical curves) and plane bending (leftward, rightward, and flat curves). Therefore, from the overall view, a road route is essentially a three-dimensional curve. The highway engineering drawings are composed of engineering drawings displaying the whole conditions of the routine and engineering drawings of bridges, tunnels, and culverts. The route drawings mainly include the road plan, route profile, and route cross-sectional drawings.

1. Route plan

The role of a routine plan is to show the direction of the route, the plane line types (straight line and left and right corner), and terrain features along both sides of a certain area. A route plan is a horizontal projection map projected from top to bottom. In other words, it is a topographic map of surrounding areas along the road plotted with the elevation projection method.

The drawing features and contents of a road plan are introduced in the following two parts: terrain and route.

1) Terrain

(1) Scale. The proportion of a route plan is generally small. The scale is usually 1 : 500 or 1 : 1000 in urban areas, 1 : 2000 in mountain areas, and 1 : 5000 or 1 : 10000 in hilly areas and plain areas.

(2) Direction. A compass or a coordinate network should be drawn in the road plan to indicate the orientation and direction of the road in an area. The arrow of the compass points to the north direction and the compass shall be drawn with thin lines. In the azimuthal coordinate network, X-axis is regarded as the north-south direction (in the north) and Y-axis is regarded as the east-west direction. The coordinate value shall be close to the marked point. The writing direction of the coordinate value shall be parallel to the grid or in the grid extension line and the value shall be marked before the axis of the axis code. As shown in Figure 5.1.2, "X2500, Y2000" indicates that the intersection point of two vertical lines is 2500 units (m) away from the origin in the North and 2000 units (m) in the east in the coordinate network.

(3) Terrain. The topography in the plan is mainly expressed with contour lines. The height difference between contour lines in Figure 5.1.1 is 2m and a thick curve shall be drawn with the interval of every four contour lines and marked with the corresponding elevation value. According to the density of the contour lines in Figure 5.1.1, it can be seen that the terrain is relatively low in the southwest and northwest

and flat on both sides of the river.

(4) Landforms. The geomorphic features on the terrain surface in the plan view such as rivers, houses, roads, bridges, power lines, and vegetation, are plotted according to the provisions of prescribed legends. The common topographic map is shown in Table 5.1.1. According to related legends, the Hong River flows from north to south in the middle of the region with rice paddies on both sides, hillsides on dry land, and fruit trees on the banks. The eastern residential area is named Chenjiazhuang Village and the western residential area is named Seshu Village. The original country road goes along the western bank of the river and power lines are arranged along the river and cross the two villages.

(5) Level. Along the route, the level of the location shall be marked near a certain interval for elevation measurements. For example, $\bigotimes \dfrac{BM12}{17.922}$ indicates that the 12th leveling point of the route has an elevation of 17.922m.

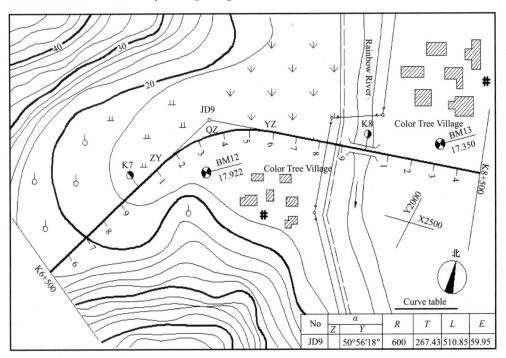

Figure 5.1.1 Route plan

Common geomorphic legends   Table 5.1.1

Names	Symbols	Names	Symbols	Names	Symbols
Houses	▨	School	文	Vegetable field	⊻ ⊻ ⊻ ⊻ ⊻
Road	-------------	Rice field	⊻ ⊻ ⊻ ⊻ ⊻	Dams	⊥⊥⊥⊥⊥

Continued

Names	Symbols	Names	Symbols	Names	Symbols
Path	 -------------	Dryland	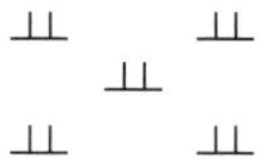 ⊥⊥ ⊥⊥ ⊥⊥ ⊥⊥ ⊥⊥	River	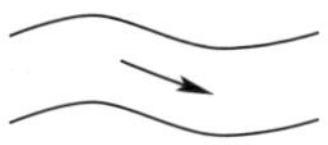 ～～→
Railway	━▬━▬━	Orchard	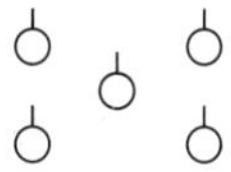 ○ ○ ○ ○ ○	Artificial excavation	 (shape with teeth)
Culvert	 ⟩———⟨	Meadow-land	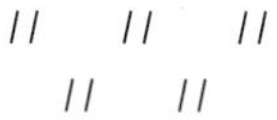 // // // // //	Low voltage power line High voltage power line	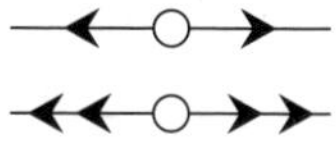 ←─○─→ ⇇─○─⇉
Bridge	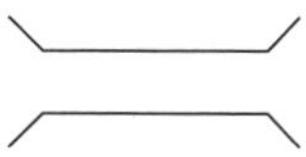 (bridge symbol)	Timberland	 ○ ○ ○ ○ ○ ○	Leveling point	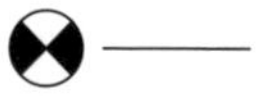 ⊗──

2) Route

(1) Designed route. A bold line is used to indicate a route because the width of a road is much smaller than its length. The width of a road can only be drawn in the plan with a larger scale. Therefore, a bold line is usually plotted along the central line of the road to represent the designed route.

(2) Mileage piles. The total length of the road route and the length between sections are denoted by the number of mileage piles. The mileage piles shall be numbered from the beginning to the end of the route and the direction of the route in the plan is always from left to right. There are two types of mileage piles: kilometer piles and 100-meter piles. The piles shall be drawn on the left side of the route. The positions of piles are denoted using symbols ⏻ and the number of kilometers is written above the sign, such as K7, indicating the distance of 7km away from the starting point. The 100-meter piles shall be marked with a short perpendicular line on the right along the routine direction. Arabic numerals are added at the end of the corresponding short line to represent the number of 100 meters and the head of Arabic numerals points to the forward direction. For example, "2" in front of the K7 kilometer pile represents the pile number of K7 +200, indicating that the pile is 7200m away from the starting point of the route.

(3) Flat curve. A route is composed of straight lines and curve segments in a plane. At the turning point in a routine, a flat curve shall be set. The most common and simple flat curve is an arc and its basic geometric elements are shown in Figure 5.1.2. JD is the intersection, the theoretical intersection of the two straight lines; α is the turning angle, the angle that the route deflects forward to the left (α_Z) or the right (α_Y); R is the curve radius, which is the radius length of the connecting arc; T is the tangent length, the length between the point of contact and intersection point; E is the external distance, the distance between the midpoint of the curve and intersection point; L is the length of the curve, the arc length between the two

tangential points of the circular curve.

In a route plan, the turning points shall be denoted with codes and numbers in turn. For example, *JD*9 indicates the ninth intersection point. The starting point of the curve segment *ZY* (straight circle), midpoint *QZ* (the midpoint of a curve), and endpoint *YZ* (straight line of a circle) shall be denoted. To indicate the geometric element values of flat curves in the route, the flat curve element table shall also be listed in the appropriate position. If a transition curve is set, then the tangential points between the transition curve and front and rear lines are respectively marked as *ZH* (straight point) and *HZ* (slow point). The tangential points between circular curves and front and rear transition curves are respectively marked as *HY* (slow dot) and *YH* (round point).

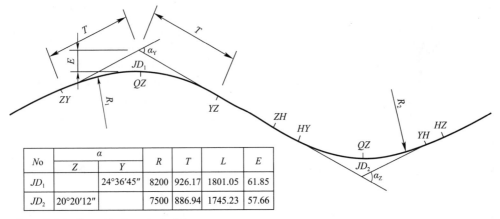

Figure 5.1.2 Geometric elements of a flat curve

No	α		R	T	L	E
	Z	Y				
JD_1		24°36′45″	8200	926.17	1801.05	61.85
JD_2	20°20′12″		7500	886.94	1745.23	57.66

As shown in Figure 5.1.1, the designed road starts from the K6+500, which is introduced from the southwest of the highland and turns right at the intersection point *JD*9 (α_Y = 50°56′18″ and circular curve radius = 600m). It goes through the north of Seshu Village, then crosses Honghe Bridge, and extends eastward.

2. Route profile

A route profile is obtained by dividing the centerline with a hypothetical vertical section and then extension, as shown in Figure 5.1.3. Since a road route is composed of straight lines and curves, the longitudinal section has both flat and curved surfaces. To indicate the route longitudinal section, the longitudinal section needs to be straightened, extended, and then drawn to obtain the route profile.

The longitudinal profile of the route mainly indicates the designed vertical lines of a road, the elevation variation along the line, the geology, and the setting of structures along with the routine.

A route profile consists of two parts: drawing and datasheet. A drawing pattern is drawn at the top of the drawing sheet and datasheets are arranged at the bottom. Figure 5.1.4 shows the vertical section of a highway from K6 to K7+600.

1) Drawing pattern

(1) Scale. The horizontal direction of the vertical section indicates the length of the route (the direction of advance), whereas the vertical direction indicates the elevation of the design line and the ground. Since the height of the route is much smaller than the length of the route, if the vertical height and the horizontal length are plotted at the same scale, it is difficult to significantly indicate the height. The general vertical scale is 10 times larger than the horizontal scale. For example, the horizontal scale in Figure 5.1.4 is 1 : 2000, whereas the vertical scale is 1 : 200. Therefore, the slope in the routine drawing is significantly larger than its actual size. To facilitate the drawing and reading, the elevation scale shall be drawn to the left of the profile.

Figure 5.1.3 Schematic diagram of the route profile

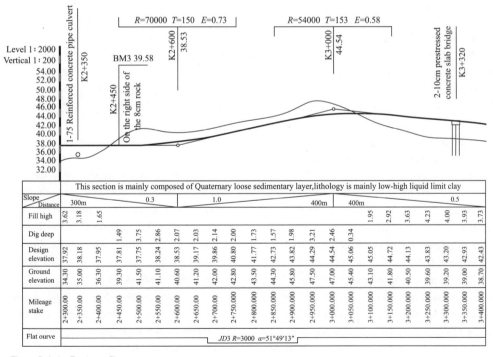

Figure 5.1.4 Route profile

(2) Design lines and ground lines. In the vertical section, the road design lines are represented by solid lines in groups, whereas the original ground line is represented by thin lines. Design lines are determined based on terrain fluctuations and highway grades according to the

corresponding engineering standards. The elevations of the points on design lines usually refer to the design elevation of the subgrade edge. Ground lines are drawn based on height measurements of center piles on the original ground. The relative position of the design line and ground line determines the excavation/filling height.

(3) Vertical curve. A design line is composed of a straight line and a vertical curve. At the transition point of longitudinal slope (variable slope point) in a design line, to facilitate the driving process of vehicles, the vertical curve of a design line shall be set according to the provisions of relevant technical standards. Vertical curves are classified into two kinds: convex and concave, which shall be respectively represented by the two symbols: ⊓ and ⊔. The central lines in the symbols shall point to variable slope points. The number of corresponding mileage piles and the midpoint elevation of the vertical curve shall be respectively marked in the left and right of the central vertical line. The two ends of the horizontal line of the two symbols shall respectively point to the starting and endpoints of the vertical curve. The values of vertical curve elements (radius, tangent length, and outer distance) are marked above the horizontal line. The number of corresponding mileage piles at the variable slope point is K2+600. The midpoint elevation of the vertical curve is 38.53m and the concave vertical curve ($R = 70000$m; $T = 150$m; $E = 0.73$m) shall be set. At the variable slope point of K3+000, a convex vertical curve ($R = 54000$m; $T = 153$m; $E = 0.58$m) shall be set.

(4) Engineering structures such as bridges and culverts along the road shall be denoted above or below the design line with vertical outgoing lines, which shall point to the center of the corresponding structure. The name, specification, and the number of corresponding mileage piles of the structure shall be denoted. For example, the center position of the culvert in Figure 5.1.4 shall be denoted by "O" to indicate single-hole tube culverts with a diameter of 75cm at the mileage pile K2+350. For example, the expression $\frac{2-10\text{m slab bridge}}{K3+320}$ indicates a prestressed concrete slab bridge with 2 spans including 10m per span at the mileage pile K3+320.

(5) Leveling point. The leveling point set along the routine should also be marked. The vertical outgoing line must point to the leveling point. The number of corresponding mileage piles and positions shall be respectively marked in the left and right of the horizontal line. The leveling point BM3 is set in a rock that is 8m away from the right side of the mileage pile K2+450 and has an elevation of 39.58m.

2) Datasheet

The datasheet of a route profile is aligned with the drawing pattern for the convenience of reading. This kind of representation method is a good way to reflect the elevation, the amount of excavation, the geological conditions, and the slope at each pile in the vertical design as well as the relationship between the flat curve and the vertical curve. The data sheet mainly includes the following items and contents:

(1) Geological profile. According to the measured data, the geological conditions along the segment shall be denoted.

(2) Slope/distance. The slope from the upper left to the upper right represents an uphill and the slope from the upper left to the lower right represents the downhill. The slope and distance are respectively marked on the upper and lower sides of the diagonal. In the datasheet in Figure 5.1.1, "0.3/300" in the first grid indicates that the route is uphill with a slope of 0.3% and a length of 300m.

(3) There are two columns in the datasheet: design elevation and ground elevation, which respectively correspond to the elevation at different points (pile numbers) of design lines and ground lines.

(4) Excavation height. When the design line is below the ground line, soil excavation is required. When it is above the ground line, soil filling is required. The excavation or filling

height shall be the absolute value of the difference between the design elevation and the ground elevation at the corresponding point (pile numbers).

(5) Numbers of mileage piles. The pile number at each point along the routine is the measured mileage value in the unit of "m" and arranged from left to right. Piles can be added at the starting point, midpoint, endpoint of the curve, and center point of the bridge.

(6) Flat curve. To denote a linear segment in a plane, a schematic flat curve is usually drawn in the datasheet. The linear segment is denoted by a horizontal line, the left turning of the road is denoted with a concave line, and the right turning is denoted with a convex line. In addition, the value of each element of the flat curve shall be sometimes denoted as well.

(7) Overelevation. To reduce the lateral force of a car on the corner of a road, the road in the flat curve should be designed according to the form of "high edge and low center". The elevation difference between the road edge and the design line is called overelevation, as shown in Figure 5.1.5.

(8) The title bar of the profile is added in the last drawing or the lower right corner of each drawing to indicate the route name and vertical and transverse scales. Denotation shall be provided in the upper right corner of each drawing to indicate the number of drawings and the total number of sheets.

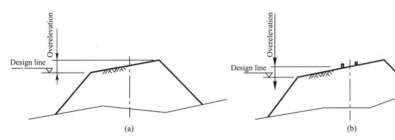

Figure 5.1.5 Overelevation of road
(a) General Road; (b) Expressway

3. Road cross-sectional drawing

The section of a route is a figure obtained from an imaginary cutting plane perpendicular to the centerline of the road. In a cross-sectional drawing, the pavement line, shoulder line, and slope line are denoted with a thick solid line. The pavement thickness is represented by a solid line, the original ground line is represented by a thin line and the center line is represented by a dotted line.

The horizontal and vertical directions of the cross-sectional drawing shall be on the same scale: 1 : 200, 1 : 100, or 1 : 50.

For subgrade construction lofting and earthwork calculation, at each center pile along the route, according to the measured data and design requirements, a series of roadbed cross-sections shall be drawn to indicate the shape of the cross-sectional of the roadbed and the ground fluctuation conditions. Pavement or road arch is generally not drawn in subgrade cross-sectional diagram. The elevation of the roadbed edge is regarded as the design elevation at the center.

There are three basic forms of roadbed cross-section:

1) Filled embankment. As shown in Figure 5.1.6(a), the whole subgrade is the filling area and called embankment. The filling height is equal to the design elevation minus the road elevation. The embankment slope is generally 1 : 1.50. The numbers of mileage piles, the embankment height of the centerline h_T and the filling area of the cross-section A_T are denoted in Figure 5.1.6.

2) Excavated subgrade. As shown in Figure 5.1.6(b), the whole subgrade is entirely excavated and called cutting. The excavation depth is equal to the ground elevation minus the design elevation and the slope excavation is generally 1 : 1. The numbers of mileage piles of

the section, the excavation height of the centerline h_W, and the excavation area of the cross-section A_W are denoted in Figure 5.1.6.

3) Half-filling and half-excavation subgrade. As shown in Figure 5.1.6(c), one part of the subgrade section is the filling area and the other part is the excavation area. The subgrade is the composite product of the two subgrades, the numbers of mileage piles, the filled (or excavation height) height of the centerline h_T, the filling area A_T, and the excavation area A_W of the cross-section are denoted in Figure 5.1.6.

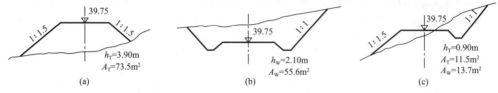

Figure 5.1.6 Basic forms of roadbed sections

4. Drawing and reading methods of road route engineering drawings

1) Drawing method of a road plan

(1) Notes for drawing a route plan drawing:

① The topographic map is drawn first. Thick contour lines are drawn before thin contour lines by hand to obtain smooth lines.

② The centerline of the route is drawn then. The centerline of the road route is drawn with the drawing instrument. Curves are drawn with bold lines ($2b$) before straight lines to distinguish them from contour lines. In **Drawing Standards of Road Engineering** (GB 50162—92), design lines and ground lines are respectively drawn with bold thick lines and bold dotted lines for comparison.

③ The road plan shall be drawn from left to right, and the mileage piles are numbered from left to right.

④ Plant legends in the plan diagram shall be drawn upward or northward. The number of drawings and the total number of sheets shall be denoted in the upper right corner of each drawing.

⑤ Plan drawings are spliced finally. As a road is long, it is impossible to draw the entire route plan in the same drawing sheet. The routine plan is usually drawn on several drawing sheets. Then, all the drawing sheets are sliced together. The number of drawings and the total number of sheets shall be denoted in the upper right corner of each drawing. The route shall be divided into segments at the mileage piles of integral miles. At both the disconnected ends, a dotted line perpendicular to the route shall be drawn as the splicing line. When adjacent drawings are spliced, the route centers shall be aligned and the splicing lines shall be coincident with each other with the north direction as the basis, as shown in Figure 5.1.7.

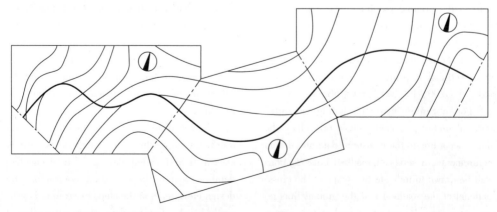

Figure 5.1.7 Splicing method of the road route plan

(2) Reading method of a plan drawing

① Control points, coordinate net (or compass direction), and scale in the route plan drawing shall be read first.

② Then the topographic map is read to know the terrain and the distribution features of ground structures.

③ The route design drawing is read then to know the direction of the road route.

④ The layout situation and elements of the flat curve are grasped.

⑤ The intersection points of the routes, highways, railways, and rivers are grasped.

⑥ All the route drawings are spliced together to know the overall layout situation of the road route in the plan drawing.

2) Drawing method of a road route profile

(1) Notes for drawing a road route profile

① Abscissae and ordinates are drawn first. On the left, ordinates indicate the data of the elevation meter and abscissae represent the data of mileage piles.

② Scale. In a vertical section drawing, the vertical scale is 10 times the horizontal scale. For example, the vertical scale is 1 : 10 and the horizontal scale is 1 : 100. The ratio of vertical and horizontal scales is generally denoted in the first diagram.

③ Ground line. A ground line is the intersection line between the cut section and the original ground. When plotting, the ground elevation points of each mileage pile are firstly dotted in the drawing pattern and then connected with thin polylines to obtain the ground line.

④ Design line of the slope. The design line is the intersection line between the profile and the designed road. When drawing, the design elevation point at each pile of the mileage shall be firstly dotted in the drawing pattern, and then solid lines are used for pulling slope so that the design line is obtained.

⑤ Line type. Thin lines and thick solid lines are respectively used to draw the ground line and design line. Mileage piles are numbered from left to right.

⑥ Variable slope point. When the slope of the road route changes, the variable slope point shall be represented by a circle with a medium diameter of 2mm. The tangent shall be represented by fine lines, whereas the vertical curve shall be represented by solid lines, as shown in Figure 5.1.8.

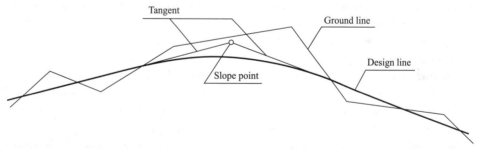

Figure 5.1.8 Vertical curve of a road

(2) Reading method of a plan drawing

① Firstly, horizontal and vertical scales and the locations of leveling points are grasped from the plan.

② Then, the ground lines are acquired to know the vertical fluctuation situation and soil distribution along the route.

③ Design lines are acquired to understand the longitudinal distribution along the route and find out the position and length of slopes.

④ Ground lines are compared with design lines to know the filling and excavation situation along the road route.

⑤ The locations of vertical curves and the index data of curve elements are grasped.

⑥ The distribution and main contents of other engineering structures along the longitudinal direction of the road route are acquired.

⑦ The correspondence between the vertical curve and the flat curve is acquired.

3) Drawing method of a road route cross-sectional profile

(1) The cross-sectional view of the subgrade shows the center piles perpendicular to the centerline of the road route. The cross-sectional drawing shall be drawn according to the following requirements:

① Thin solid lines are used to draw ground lines of cross-sectional profile. Thick solid lines are used to draw design lines. Overelevation and widening of the road shall also be shown.

② The cross-sectional diagram of subgrade drawn in the same drawing shall be arranged by the number of the mileage piles from the bottom left to the top right.

③ In the upper right corner of each cross-sectional diagram of the subgrade, the drawing number and the total number of sheets shall be provided and the icon shall be drawn in the bottom right corner of the last drawing.

(2) Reading method of a plan drawing

A cross-sectional diagram of the subgrade is generally read along with the piles from bottom to top and from left to right to grasp the elevation of the subgrade, the subgrade slope, embankment height, filling/excavation height, and filling/excavation area at each pile.

5.2 Urban Road Route Drawings

Urban roads are generally composed of roadway, sidewalk, green belt, median lane, crossing, traffic plaza, viaduct highway, underground road, and other facilities. A typical urban road cross-sectional layout is often referred to as the three-plate form with the wide two-way driving motorway in the center, one-way non-motor vehicle lanes on both sides, and the outermost sidewalk. The one-way non-motor vehicle lanes are separated by the green belt.

Like road route drawings, the linear design drawings of urban roads are also expressed in the forms of plans, longitudinal sections, and cross-sectional views. Since the terrain of urban roads is generally flat and the design of urban roads is based on the implementation of city planning and traffic planning, the traffic properties and components of urban roads are much more complex than those of highways. Therefore, the cross-sectional drawings of urban roads are much more complex than those of highways.

The cross-sectional view of an urban road is a cross-sectional view in the normal direction of the road centerline. The urban road cross-section is composed of roadway, sidewalk, green belt, and median lane. In a city, the area between the red lines on two sides of the street is used for constructing urban roads.

1. Cross-section drawings

1) Basic forms of urban road cross-sectional layout

According to the different layouts of motorway and non-motorized lanes, the road cross-section is arranged in the following four basic forms:

(1) One-board section. All the vehicles are organized on the same carriageway. Motor vehicles are organized in the middle, whereas non-motor vehicles are arranged on both sides, as shown in Figure 5.2.1(a).

(2) Two-plate section. It is separated from the center of a road by a median lane or median piers so that the whole traffic is divided into two directions, but the traffic in the same direction is still mixed, as shown in Figure 5.2.1(b).

(3) Three-board section. Motor vehicles and non-motor vehicles are separated from each other by two median lanes or median piers. The vehicle way is divided into three parts: the middle two-way lane, and two one-way non-motor vehicle lanes with opposite traffic on both sides,

as shown in Figure 5.2.1(c).

(4) Four-board section. Based on the three-plate section, a central belt is added to facilitate the motor vehicle to drive, as shown in Figure 5.2.1(d).

2) Contents of a cross-sectional drawing

The final results of the cross-sectional design are shown in standard cross-sectional drawings. Various components of the cross-section and their relationships are denoted in the drawings. Figure 5.2.2 shows the cross-sectional view of a recent design. To clearly show the height difference, the scale in the height direction (longitudinal) is 1 : 50, whereas the scale in the horizontal direction (horizontal) is 1 : 200.

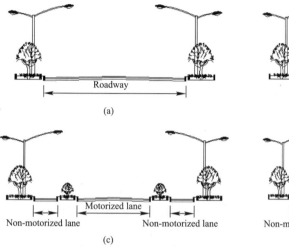

Figure 5.2.1 Road cross-sectional layout

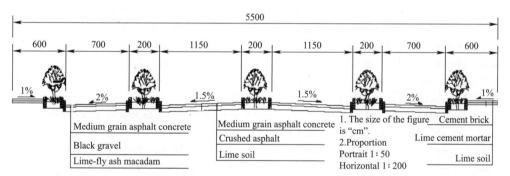

Figure 5.2.2 Cross-sectional drawing of urban roads

As shown in Figure 5.2.2, the road adopts a four-plate section facilitating motor vehicles and non-motor vehicles to respectively run in a single-way lane. Sidewalks are arranged on both sides and a median lane is arranged in the middle. The width of each component and structural design requirements are also provided in Figure 5.2.2.

In addition to the recently designed cross-sectional drawing, a cross-sectional drawing of the long-term planning and design shall be plotted for road construction in stages. To design the amount of earthwork and construction layout, the centerline cross-section is given in the same way as the road cross-section. The designed cross-sectional drawing shall be plotted and the numbers of mileage piles and design elevation shall be provided.

2. Plan drawings

Similar to highway route plan, an urban road plan is used to indicate the direction of the urban road, line types, the roadway layout, and the terrain and geographical situation on both sides of the roadway within a certain range.

Figure 5.2.3 is a plan drawing for a segment of an urban road, Nanping Road. It mainly represents the graphic design of crossings and urban roads. The contents of the urban road plan drawing can be classified into two parts: roads and terrain.

1) Road conditions

(1) The road centerline is indicated by a dash-dotted line. To indicate the road length, the mileage is added on the centerline of the road. As shown in Figure 5.2.3, the plan drawing shows a road segment from 0+000 to 0+330.95.

(2) The direction of the road is determined by the coordinate network (or the compass). It can be seen from the compass direction that the direction of road goes from east to west.

(3) The scale of an urban road plan drawing is larger than that of the road route drawing, so the distribution sizes and widths of the roadways and sidewalks can be drawn proportionately. As shown in Figure 5.2.3, roadways on two sides are 6m wide and non-motorways are 4m wide. In addition, dumm drifts are arranged on both sides of the road.

2) Terrain and geographical situation

(1) The topography of urban roads is generally flat and the terrain is represented by a large number of topographic points as well as contour lines.

The design end of JianGong East Street is 0+330.95.

(2) This road segment is a new urban road in the suburbs. The new road mainly passes through dry land and occupies some houses. The terrain and geomorphology of the area can be found in Table 5.1.1 and Table 5.2.1.

3. Vertical section diagram

The vertical section diagram of an urban road is also a cross-sectional diagram along the centerline of the road. Its function is the same as that of the highway route profile and its contents are composed of two parts: drawing patterns and datasheets.

1) Pattern part

The pattern part of the urban road profile is the same as that of the road route profile. In addition to the ground line of the road centerline and longitudinal slope design line, some other elements such as the vertical curve, the construction height, the location along the bridge, structure type, aperture, the locations and elevations of crossings, the position of the leveling points along the line, and numbers and the elevations of piles. A vertical section diagram is drawn in the Cartesian coordinate system at an abscissa scale of 1 : 500 to 1 : 1000 and an ordinate scale of 1 : 100 to 1 : 50. That is, the vertical scale is 10 times the horizontal scale.

In addition, in the longitudinal section design drawings of an urban road, the names of crossing roads, the foot elevation of the crossings, and the elevation of entrances and exits of neighborhood and important buildings.

2) Datasheets

The section profile of an urban road is the same as the longitudinal section drawing of the highway route. The section profile only corresponds to the pattern part and relevant design contents shall be denoted. In the columns of datasheets, from bottom to top, the following items shall be filled: straight line and flat curve diagrams, the number of mileage piles, ground elevation, design elevation, filling height, slope/slope length, address, drainage ditch ground line, elevation, flow direction, and other relevant contents.

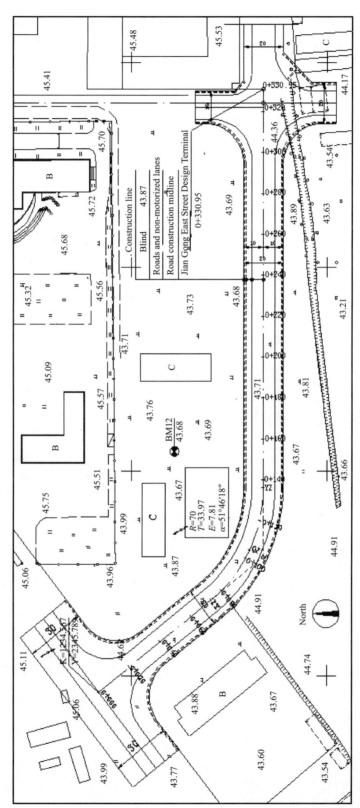

Figure 5.2.3　Jian Gong East Street construction plan

Legends of common ground objects Table 5.2.1

Names	Symbols	Names	Symbols	Names	Symbols
Simple house with only roof		Asbestos	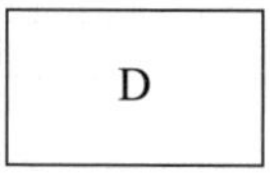 D	Sanitary sewer manhole	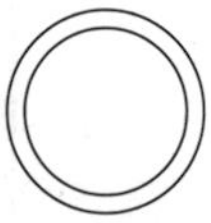
Masonry or concrete structures	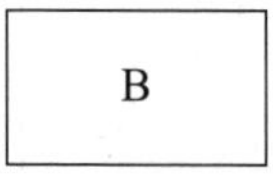 B	Wall	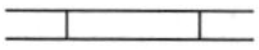	Communi-cation rod	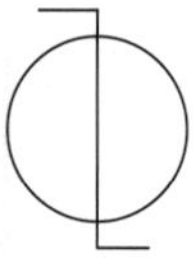
Brick house	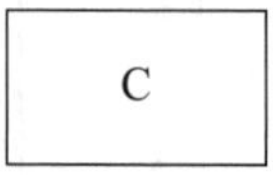 C	Storage tank	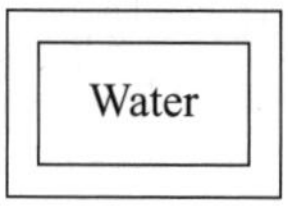 Water	Fence bar	

Chapter 6
Bridge Engineering Drawings

A road may cross rivers, lakes, mountains, and another road. When a road intersects with other routes (such as railways), to keep the route smooth, it is necessary to build bridges, which can ensure the traffic on the bridge, water discharge below the bridge, the normal passage of vessels navigation or road and railway traffic under the bridge. A bridge is an important part of road engineering, as shown in Figure 6.1.1.

Figure 6.1.1　Bridge engineering

6.1　Bridge Engineering Drawings

1. Overview

1) Basic components

As shown in Figure 6.1.2, a bridge consists of three parts: the bridge span structure (beam/arch ring and bridge deck system, also called superstructure), substructure (abutment, pier, and foundation), and subsidiary structure (railing, lamp post, revetment, diversion structure, etc.)

The bridge span structure is the main load-bearing structure crossing a barrier that interrupts a route and it is also called a superstructure.

Piers and abutments are structures that support the bridge span structure and transfer loads such as dead loads and vehicle loads to the foundation known as the substructure.

The support is a force transmission structure designed to connect the bridge span structure and the piers or abutments of the bridge.

At the joint between the embankment and abutment, the stone taper revetment is generally set on both sides to ensure the stability of the embankment slope.

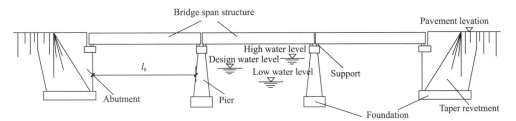

Figure 6.1.2 Basic components of a bridge

The water level is variable. The lowest water level in the dry season is called the low water level, whereas the highest water level in the flood season is called the high water level. The water level used in the bridge design is calculated based on the high water level and specified flood frequency also called design water level.

The net span (l_0) is the net distance between two adjacent piers at the design water level.

The total span (l) is the sum of the net spans of the bridge and reflects the ability of the bridge to discharge water.

The length of the bridge (bridge length L) is the distance between the sidewalls of the two abutments at both ends or the rear end of the taper revetment. For a bridge without abutment, the length of the bridge refers to the full length of the lane.

2) Classification of bridges

Bridges exist in many forms and are commonly classified according to the following ways:

(1) According to the structural style, bridges can be classified as beam bridges, arch bridges, rigid frame bridges, suspension bridges, cable-stayed bridges, and so on.

(2) According to building materials, bridges can be classified as steel bridges, reinforced concrete bridges, stone bridges, wooden bridges, and so on. Among them, reinforced concrete beam bridges are the most widely used.

(3) According to the length and span, bridges can be classified as grand bridges, great bridges, medium bridges, and small bridges (Table 6.1.1).

Bridge classification Table 6.1.1

Bridge classification	Multiple-span bridge (m)	Single-hole span (m)	Bridge classification	Multiple-span bridge (m)	Single-hole span (m)
Grand bridge	$L \geqslant 500$	$L \geqslant 100$	Medium bridge	$30 < L < 100$	$20 \leqslant L \leqslant 40$
Great bridge	$L \geqslant 100$	$L \geqslant 40$	Small bridge	$8 < L < 30$	$5 \leqslant L \leqslant 20$

(4) According to the position of the superstructure, bridges can be classified as deck bridges, through bridges, and half-through arch bridges. A bridge arranged above the main load-bearing structure is called the deck bridge; a bridge arranged under the main load-bearing structure is called the through bridge; a bridge arranged in the middle of the main load-bearing structure is called the half-through bridge, as shown in Figure 6.1.3.

A bridge should meet the requirements of function, economics, artistics, construction, and so on. Before the construction, it is necessary to first survey the topography, geology, hydrology, building materials, and other aspects related to the bridge and then the topographic drawing and geological sections are drawn for the architectural and structural construction design. The bridge design is generally divided into two stages: the preliminary design stage and the

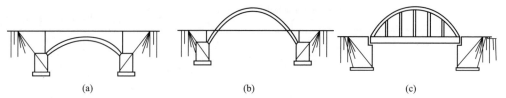

Figure 6.1.3　Bridges are classified according to the position of the superstructure
(a) Deck bridge;(b) Half-through bridge;(c) Through bridge

construction design stage.

Although the structural forms and building materials of bridges are different, the drawing method is the same. The bridge engineering drawings can be generally classified into site plan, geologic section drawing, construction drawings, component drawings, detail drawings, and so on. Here, taking a reinforced concrete beam bridge as an example, the contents and characteristics of bridge engineering drawings are introduced below.

2. Bridge engineering drawings of reinforced concrete beam bridges

1) Site plan

The site plan of a bridge mainly shows the location, its connection with the route, as well as terrain and ground features. The drawing method is the same as that of the route plan, but the scale is relatively large. Through the topographic survey, the road, river, leveling point, drilling holes, and surrounding terrain and ground features are plotted in the bridge site plan to provide the basis for the bridge design and construction positioning.

Figure 6.1.4 shows the site plan of Hongqiao Bridge. In addition to the route, topography, terrain, and surface features, the drilling location, mileage, and benchmark shall be denoted. The standard symbols of vegetation should be arranged in the north direction. The texts can be written based on the route requirements and the direction in the general drawing.

2) Geological section drawing

The geological section drawing of a bridge is the river bed geological section drawing, which is drawn based on the data obtained from the hydrological survey and geological drilling and shows the location of the geological and hydrological condition of the bridge. It includes the river bed section contour, the high water level line, the design water level line, and the low water level line. It can be regarded as the basis for the design of bridges. As for small bridges, it is not necessary to draw the geological section and only the geological descriptions shall be provided. To show the geological conditions and river bed depth, the scale of terrain height (elevation) and the horizontal scale shall be provided. As shown in Figure 6.1.5, the terrain height scale is 1 : 200 and the horizontal scale is 1 : 500.

3) Construction drawings

Construction drawings of a bridge show the main layout and components of the bridge. They mainly show the type, path, number of spans, the overall size, elevation, the bridge width, the relationship between the main components, the elevation of each part, materials, general technical practices, etc. They may be used as the basis for locating piers, installing components, and controlling elevation during the construction. Construction drawings of a bridge generally include the plan, elevation, and section drawings.

Figures 6.1.6 – Figure 6.1.8 are the construction drawings of the Baisha River Bridge with a scale of 1 : 200. The bridge is a three-span reinforced concrete simply supported beam bridge. The total length is 34.90m and the total width is 14m. The middle span is 13m long and the side span is 10m long. The bridge has two pillar piers and two ends are the gravity-type concrete abutment. The foundation of the abutment and piers are reinforced concrete precast driven piles. The superstructure of the bridge is a reinforced concrete hollow slab beam.

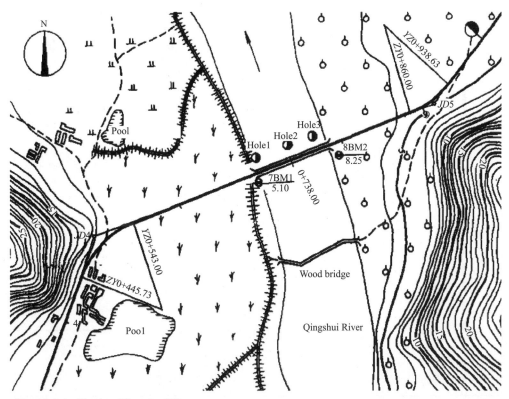

Figure 6.1.4 Site plan of Hongqiao Bridge

(1) Plan

The plan of a bridge is in the form of a semi-profile. The left half gives the top view of the bridge which mainly contains the roadway, sidewalks, railings, and other places. As shown in Figure 6.1.6, the width of the roadway is 10m and the width of sidewalks on both sides is 2m. In the right half, with the cutaway drawing method (or layered painting), assuming that the superstructure is removed, the plan shape and location of the No. 2 pier and the right side of the abutment are shown. The dashed circle in the pier is the projection of piles and the dashed square in the abutment is the projection of the lower piles.

(2) Elevation

The bridges are generally symmetrical, so the elevation drawings are often synthesized with semi-elevation and semi-longitudinal section drawings. As shown in Figure 6.1.7, the left half shows the external view of the left side abutment, the No. 1 pier, slab beam, sidewalk railings, and, other major parts. The right half is obtained by longitudinally cutting along the bridge center line and shows the No. 2 pier, the right-side abutment, beam, and bridge deck. The elevation also shows the cross-section shape of the river bed and the structures below the river bed (abutment and piles), which shall be drawn by the dashed line in the semi-elevation drawing and drawn in solid line in the semi-longitudinal section drawing.

Since the precast-driven piles are long, it is not necessary to draw the whole length of piles and the break lines are adopted in the middle. The elevation drawing also shows the important positions such as bridge deck, bottom beam, pier, abutment, a top elevation of piles, and normal water level (average annual level).

(3) Section

As indicated in the elevation drawing, the I-I section is cut at the mid-span position and the II-II section is cut at the side-span position. The section of the bridge is the combination of the left half I-I section and the right half II-II

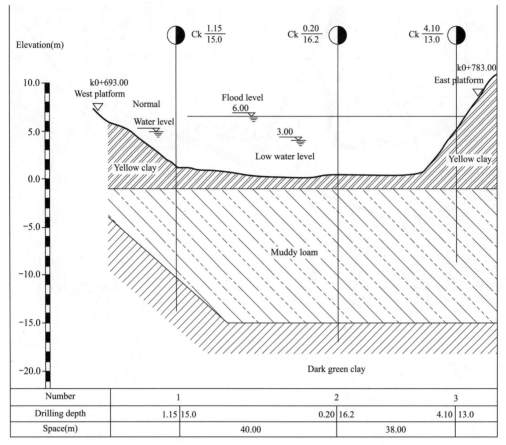

Figure 6.1.5 Geological section drawing of a bridge

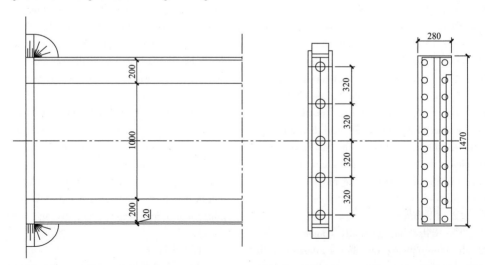

Figure 6.1.6 Plan of a bridge (cm)

section (Figure 6.1.8). The superstructures of the middle span and side span are the same. The total width of the bridge deck is 14m, which is composed of a 10m wide reinforced concrete hollow slab beam and two 2m wide sidewalks on both sides. The material symbol is

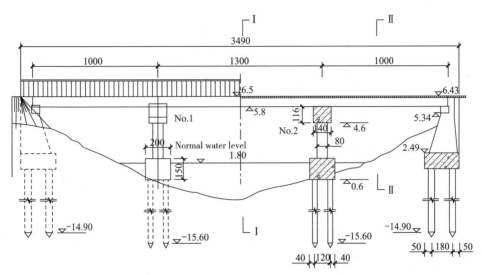

Figure 6.1.7 Elevation of a bridge (cm)

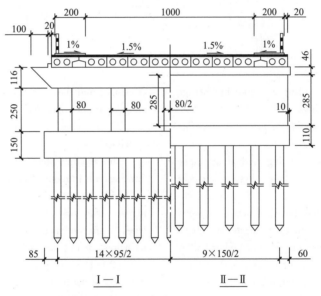

Figure 6.1.8 Section of a bridge (cm)

not shown in this section drawing as the drawing is too small to show the material symbol. The left half I - I section shows the components of the piers, including the pier cap, piers, bearing platform, and piles. The right half of the II - II section shows the components of the abutment, including the platform cap, platform body, piles, etc.

4) Component drawings

Figure 6.1.9 is the component drawing of a bridge.

Since the scale of construction drawings of a bridge is small, it is impossible to show various components of the bridge in detail. Therefore, according to actual construction demands, the shapes and sizes of each component and steel structures shall be drawn on a larger scale, which is commonly 1 : 50–1 : 10. Some local detail drawings can be drawn in larger scales, such as 1 : 3 and 1 : 10. The component drawing method is the same as that of local detail drawings.

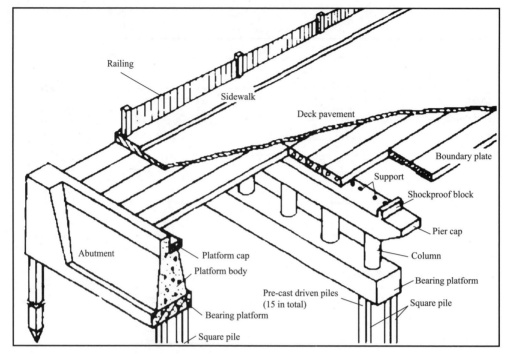

Figure 6.1.9 Component drawing of a bridge

3. Steps of reading and plotting bridge engineering drawings

1) Reading method of bridge engineering drawings

(1) The basic reading method of bridge engineering drawings is the physical analysis method. Although a bridge is a large and complex structure and composed of many necessary members, it is not difficult to master the shape and size of the whole bridge after understanding the site plan, the overall layout, each component's drawing, and the relationship between members.

(2) The drawings are read repeatedly from the overall layout to the details and then from the details to the overall layout. The bridge engineering drawings shall be read from a large scale to a small scale and from complex structures to simple structures.

(3) According to the projection relationships, the drawings are compared to have a clear idea of the whole bridge. Projection drawings cannot be read alone. Projection drawings shall be read together with other projection drawings including the site plan and detail drawings, and the lists of concrete, reinforcing bars, and other materials.

2) Reading steps of bridge engineering drawings

The reading steps of bridge engineering drawings are introduced below:

(1) The title and illustrations of the drawing are read first to grasp the bridge name, type, main technical indicators, construction measures, scale, size units, etc.

(2) The site plan and geological section drawings of the bridge are read then to know the location, hydrology, and geological conditions of the bridge.

(3) Bridge construction drawings are read then to master the bridge type, span number, size, total length, the total number of piers, and know the riverbed section and geological conditions. It is necessary to read the elevation drawings (including the longitudinal section) and compare them with the plan to know the width of the bridge, the size of sideways, and the form of the main beam section. In a section drawing, it is necessary to find out the cutting line and observation direction and have a pre-

liminary understanding of the bridge.

(4) The component drawings and detail drawings are respectively read to find out the detailed structure of the components. After reading component drawings, the general drawings are read again to understand the configuration and size of various components until all of them are grasped.

(5) The construction materials shall be grasped after reading the bridge drawings. The material quantity list of the project, the detailed list, and explanation of the concrete, reinforcing bars and other materials shall be read again to check for size errors or omissions.

The drawing method of bridge engineering drawings is the same as that of other engineering structures. The number, scale and sheet size of drawings (including section and cross-section drawings) shall be determined firstly, as listed in Table 6.1.2.

Common Scales Table 6.1.2

Items	Names of drawings	Descriptions	Scales	
			Common scales	Classification
1	Site plan	To illustrate the location of a bridge, routes, topography, and surrounding terrains. Bridges, houses, and crops are only denoted with schematic symbols	1:500–1:2000	Small scale
2	Geological section	To indicate the river bed, geological section, and hydrological conditions at the bridge site. To highlight the river-bed fluctuation, the scale in the height direction should be several times larger than that in the horizontal direction	1:500–1:100 (height direction) 1:2000–1:500 (horizontal direction)	Ordinary scale
3	Bridge construction drawings	To show the overall view, length, height, size, and navigation of the bridge and the locations of all components. The scale of the section can be enlarged 1 to 2 times larger than that of the elevation	1:50–1:500	
4	Component drawings	To indicate the structures of beams, abutments, sidewalks, and railings	1:10–1:50	Large scale
5	Detail drawings	To show the bending and welding of concrete and reinforcing bars, the carvings of the railings, details, etc.	1:10~1:3	Large scale

Drawing steps:

(1) Drawing the baselines of each projection drawing

According to the selected scale and the relative projection position, the drawings shall be evenly arranged in the drawing frame. It is necessary to retain enough space for icons, descriptions, projection titles, and dimensions during the arrangement. When a projection position is determined, the centerline of each projection drawing is selected as the baseline and drawn as the start step.

(2) Drawing the main contours of each component

Taking the baselines as the guidelines, the main contours of each component are drawn according to the elevation and dimensions of each component.

(3) Drawing the details of each component

All the components are drawn according to the order from large components to small components based on main contours. The corresponding lines of each projection drawing shall be aligned with each other. The section, railings, slope symbols, elevation symbols, and dimension lines are then drawn step by step.

(4) Deepening or linking thick lines

After checking the details, thick lines can be deepened or linked. Finally, dimension annotation and other items are added.

6.2 Tunnel Engineering Drawings

A tunnel is constructed when a road crosses a mountain. Tunnel engineering involves the tunnel body, lining structures, entrance, and accessory structures. The drawing method of tunnel engineering is the same regardless of the structure and material. Although a tunnel is long, the shape of the section is rarely changed. Therefore, in addition to the site plan of a tunnel, the tunnel construction drawings are mainly composed of tunnel entrance drawings, cross-section drawings (the shape of tunnel and lining), and refuge hole drawings.

1. Tunnel entrance drawings

Tunnel entrances can be classified as end wall type, wing wall type, and column type. Figure 6.2.1 shows the three-dimensional view of the three types of tunnel entrances.

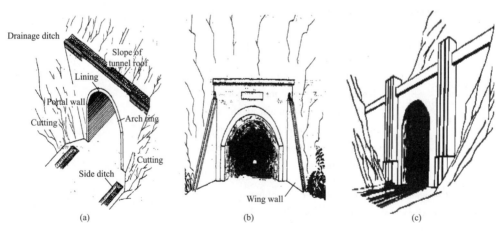

Figure 6.2.1 Three-dimensional views of three tunnel entrances
(a) End wall type;(b) Wing wall type;(c) Column type

1) Elevation drawing

An elevation drawing is the facade projection of the tunnel entrance. As for symmetrical or unsymmetrical entrances, the whole entrance shall be drawn. The elevation drawing shall reflect the style of the entrance wall. The top part above the entrance wall is the top cap. It also shows the lining section of the tunnel and the lining section is composed of an arched wall ($R=400$cm) with a thickness of 45cm and two straight sidewalls. The height of the tunnel entrance is 740cm and the length is 790cm. There is a dotted line tilting from left to right above the entrance wall and an arrow denoted with $i=$

0.02, indicating that there is a drainage ditch with a slope of 2% at the top of the entrance. The arrow points to the direction of water flow. The other dotted lines indicate the invisible contours of the entrance wall and the bottom of the tunnel, which are covered by the cutting slope on both sides of the entrance and the road surface.

2) Plan drawing

Only the projection of the exposed part of the entrance is drawn in the plan. The plan drawing shows the width of the top cap of the entrance wall, the structure of the drainage ditch on the roof, and the position of the side ditch on both sides (the side ditch section is not shown).

3) Section drawing

The section drawing of the entrance only shows a short segment from the entrance into the tunnel. The inclined slope of the entrance wall is 10 : 1 in the I-I section in Figure 6.2.2 and the thickness of the entrance wall is 60cm. It also shows the section shape of the ditch, arch circumference thickness, and materials of the cross-section. For the convenience of reading, numbers are used in Figure 6.2.2 to indicate different components on the three projections, respectively. For example, the entrance wall is denoted with ①′, ①, ①″; the drainage ditch of the cave roof is denoted with ②′, ②, ②″; the arch is denoted with ③′, ③, ③″; the top cap is denoted with ④′, ④, ④″.

2. Refuge hole drawings

There are two kinds of refuge holes: a large one and a small one. They are set to keep pedestrians, tunnel maintainers, and trolleys away from vehicles. They are alternatively arranged on the sides of the tunnel along the direction of the road. Usually, one small refuge hole is set every 30m and one large refuge hole is set every 150m. To show the layout and positions of the refuge holes, longitudinal sections and plan drawings of the tunnel are needed.

As shown in Figure 6.2.3, the content of the plan is simple. To show the information effectively on one drawing sheet, vertical and horizontal directions can be drawn in different scales. The scale of vertical direction is often 1 : 2000 and the scale of horizontal is often 1 : 200.

1) Longitudinal section drawing

The longitudinal section drawing shows the shape and position of large and small refuge holes. It reflects the lining materials of the tunnel vault and the profile of the tunnel.

2) Refuge hole plan drawing

The plan drawing mainly shows the depth and shape of refuge holes, including large and small ones, and the overall layout of refuge holes in the tunnel.

3. Culvert

1) Definition of the culvert

A culvert is the structure constructed where the highway or railway intersects with ditches or roads. It can also be a small structure that water, pedestrian flow, and traffic flow passing through. A culvert is filled with earth, whereas a bridge is paved directly on the bridge span structure (but there is still a ballast). A culvert is like a hole dug in the roadbed, whereas the roadbed is broken off at the bridge. Culvert and bridge are technically distinguished from each other based on the standards of span. In general, the standard span L_0 of a single-hole culvert is less than 5m and the total length of a multiple-span culvert is less than 8m. A small bridge with a length of 5m to 20m is usually classified as a kind of culvert due to its simple structure. However, pipe culverts and box culverts, regardless of the size, are called culverts.

2) Composition and characteristics of culverts

A culvert is mainly composed of three parts: cave, foundation, and entrances. The main function of the culvert is to bear live loads and earth pressure. The role of culvert entrances is to prevent the culvert foundation and the roadbed on both sides from erosion and ensure smooth water flow.

A culvert has the following characteristics:

(1) It has enough capacity to discharge the flood and ensure that the flood (in case of the flood with a return period of 50 years) can

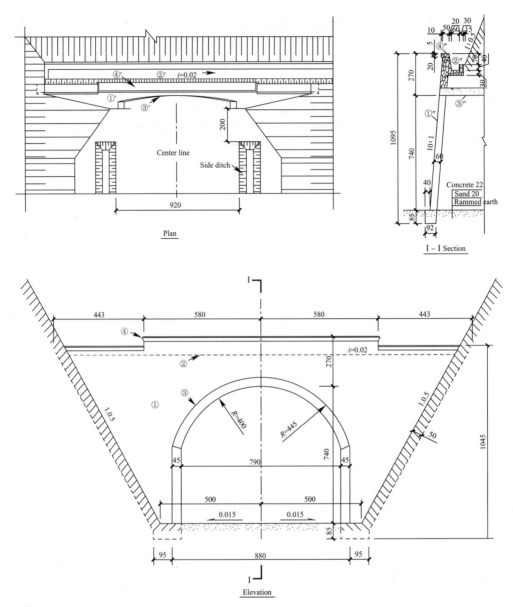

Figure 6.2.2 Tunnel entrance drawings

be discharged smoothly and quickly.

(2) It has sufficient strength and stability to ensure the displacement and deformation of the components inside the required limits.

(3) It has high reliability and durability to ensure a certain service life in the natural environment.

3) Classification of culverts

According to building materials, culverts can be classified as stone culverts, brick culverts, concrete culverts, and reinforced concrete culverts. According to the situation of earth filling on the cave roof, culverts can be classified as open culverts without earth filling on the cave roof and buried culverts with the earth filling thickness of more than 50cm on the cave roof. According to the cross-sectional shape, culverts can be classified as circular culverts, oval culverts, arched culverts, trapezoidal culverts, and rectangular culverts. According to the number of holes, culverts can be classified as single-hole culverts, double-hole culverts, and multiple-hole culverts. According to the structure form, culverts can be classified as pipe culverts, slab culverts, arch culverts, and box culverts, as listed below.

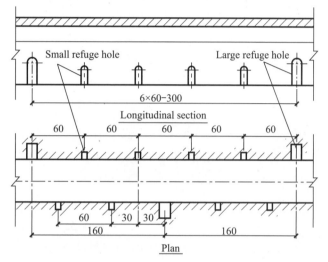

Figure 6.2.3 Refuge hole plan (longitudinal 1:20; horizontal 1:200)

(1) Pipe culverts

A pipe culvert is mainly composed of the pipe body, foundation, seam, and waterproof layer, as shown in Figure 6.2.4.

Figure 6.2.4 A pipe culvert

(2) Slab culverts

Slab culvert is mainly composed of cover, culvert, tunnel bedding, expansion joints, waterproof layer, and so on, as shown in Figure 6.2.5.

(3) Arch culverts

An arch culvert is mainly composed of the arch ring, back haunch fillet, foundation, bedding, settlement joints, and drainage facilities, as shown in Figure 6.2.6.

Figure 6.2.5 A slab culvert

(4) Box culverts

A box culvert is mainly composed of the reinforced concrete culvert, wing wall, foundation, deformation joint, and so on, as shown in Figure 6.2.7.

4) Forms of culvert entrances

Culvert entrances can be sorted as eight-character type, straight wall type, cone slope type, end wall type, and so on. The river bed must be paved regardless of the form of entrances, as shown in Figure 6.2.8.

(1) Eight-character type

Figure 6.2.6 An arch culvert

Figure 6.2.7 A box culvert

It is open inclined and the height of both sides of the eight-shaped wing wall varies with the slope of the embankment. The eight-character entrance has the advantages of less engineering work, good hydraulic performance, simple construction, and low cost, so it is the most common entrance type.

(2) Straight wall type

It can be regarded as an open mouth for the zero angles of the octagonal cave. It is suitable for a narrow and deep cave and the depth is the same as the width of the culvert. Since the wing wall is short and the cost is low, it is applicable for the configuration of the stilling basin.

(3) Cone type

Based on the end wall type, it is paved on the surface of cone-shape filling of the lateral protrusion. The masonry volume is large and uneconomical, but it is stable for higher culverts. It is a widely used form.

(4) End wall type

The structure is perpendicular to the longitudinal axis of the culvert and partially blocks the low embankment slope. The cave height is determined by the height of the backwater in front of the culvert or the height of the shoulder. The construction of an end wall type culvert is simple, but its hydraulic performance is not so good and only suitable for situations with a low flow rate and low erosion.

5) Culvert drawings

According to the characteristics (narrow and long) of culvert structures, regarding the direction of the water from left to right, the vertical profile is used for the elevation drawing. The break line is used to divide the profile into two halves, to show the entrance and exit sides separately. It is also possible to increase the section drawing perpendicular to the longitudinal direction and express it in a semi-profile form. The entrance elevation is required to complement the elevation drawing of the culvert. If the entrance and the exit of the culvert are of different shapes, it is necessary to draw the elevation drawing for each of them. The plan drawing of the culvert excludes the non-roof cover soil to make it clear. If necessary, the plan is converted into a half-section, half foundation top section drawing. In addition to the above three drawings, structural detail drawings are also required, such as steel members, wing wall section, and so on.

Taking the slab culvert as an example, the culvert drawings can be seen in Figure 6.2.9.

6) Reading the culvert drawings

Reading steps of culvert drawings:

(1) The title bar and instructions are read to grasp the type, diameter, scale, size units, materials of culverts.

Figure 6.2.8 Entrance form
(a) Eight-character type;(b) Straight wall type;(c) Cone type;(d) End wall type

(2) The drawings are read and grasped their relationships.

(3) As for various components of the culvert, their structure and their sizes are provided.

(4) Based on the above analysis, the overall shape of the culvert and the size of each part being imaged.

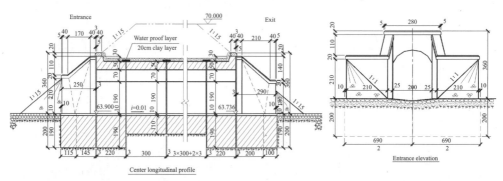

Figure 6.2.9 Plan, section, and elevation of a slab culvert (one)

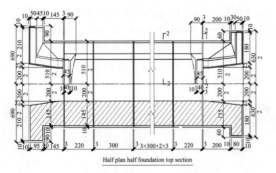 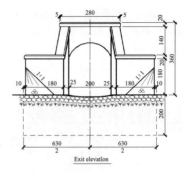

Figure 6.2.9 Plan, section, and elevation of a slab culvert(two)

Chapter 7
Computer Aid Drawings—
AutoCAD

7.1　Introducing AutoCAD

1. Opening AutoCAD

AutoCAD is designed to work in a Windows operating system. In general, to open AutoCAD, either left-click on the AutoCAD shortcut on the Windows desktop or start from the Start menu. After opening it, the Start mode page may appear, there are two pages, Learn and Create, as shown by Figure 7.1.1 and Figure 7.1.2. The Create page supplies five options of starting a drawing, creating a new drawing, opening an existed dwg. file, opening sheet set, getting more templates online, and exploring sample drawings from the computer, as shown by Figure 7.1.2.

Figure 7.1.1　Settings for page1 in the Start mode page

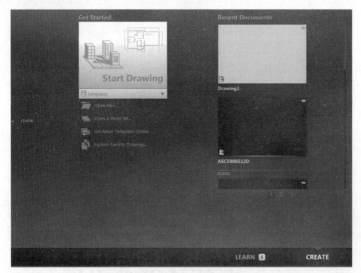

Figure 7.1.2　Settings for page1 in the Start mode page

When AutoCAD is opened, a window appears, which depending upon whether a 3D Modeling, Classic AutoCAD, or a 2D Drafting & Annotation workspace has been set as the QNEW in the Options dialog. In this example, the 2D Drafting & Annotation workspace is shown and includes the Ribbon with Tool panels (Figure 7.1.3). This workspace can be changed by the "Workspace switching" tool, which is set at the bottom status bar, as shown by Figure 7.1.4.

The 2D Drafting & Annotation workspace shows the following details:

1) **Quick Access toolbar**: The toolbar at the top right of the AutoCAD window holds several icons, including the open tool icon, save icon, return and forward ions, and print icon. For example, if you click the open tool icon, a select file dialog will appear on the screen, as shown by Figure 7.1.5.

2) **Menu Browser icon**: A left-click on the arrow to the right of the A symbol at the top left-hand corner of the AutoCAD window causes the Menu Browser menu to appear.

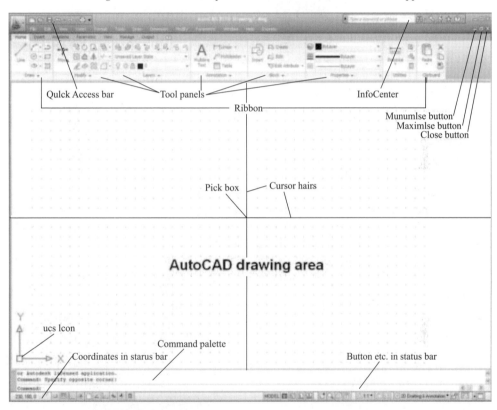

Figure 7.1.3 AutoCAD Drafting & Annotation workspace

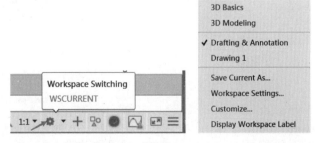

Figure 7.1.4 The Workspace Switching menu

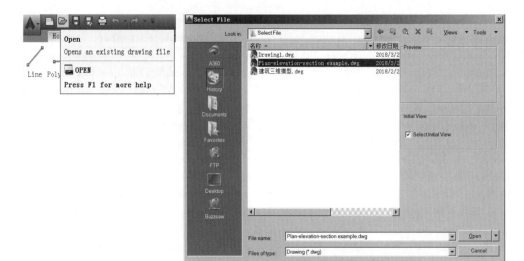

Figure 7.1.5   The open icon in the quick access toolbar brings the select file dialog on the screen

3) **Ribbon area**: Which includes tabs, each of which when clicked will bring a set of panels containing tool icons. Further tool panels can be seen by clicking the appropriate tab. The panels in the ribbon can be changed to previous AutoCAD releases using the Customer User Interface dialog if desired.

4) **Tool panels**: Each shows tools appropriate to the panel. Taking the Home/Draw panel as an example, Figure 7.1.6 shows that placing the mouse cursor on one of the tool icons in a panel brings a tooltip on screen showing details of how the tool can be used. Two types of tooltips will be seen. In the majority of future illustrations of tooltips, the smaller version will be shown. Other tools have pop-up menus appearing with a click. The example given in Figure 7.1.7, a click on the Circle tool icon will show a tooltip. A click on the arrow to the right of the tool icon brings a pop-up menu showing the construction method options available for the tool.

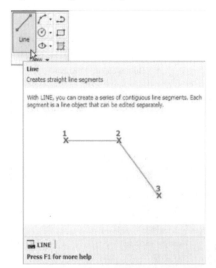

Figure 7.1.6   The descriptive tooltip appearing with a click on the Line tool icon

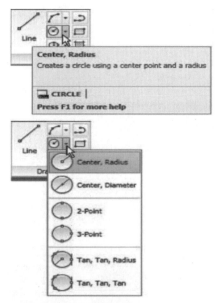

Figure 7.1.7   The tooltip for the Circle tool and its pop-up menu

2. The mouse as a digitizer

Many operators working in AutoCAD will use a two-button mouse as a digitizer. There are other digitizers which may be used——pucks with tablets, a three-button mouse, etc. Figure 7.1.8 shows a mouse which has two buttons, and a wheel.

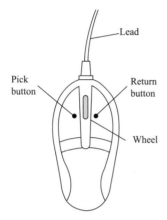

Figure 7.1.8 A two-button mouse

To operate this mouse pressing the Pick button is a left-click. Pressing the Return button is a right-click. Pressing the Return button usually, but not always, has the same result as pressing the Enter key of the keyboard.

When the wheel is pressed, drawings in the AutoCAD screen can be panned by moving the mouse. Moving the wheel forwards enlarges (zooms in) the drawing on screen. Move the wheel backwards and a drawing reduces in size. The pick box at the intersection of the cursor hairs moves with the cursor hairs in response to movements of the mouse. The AutoCAD window includes cursor hairs which stretch across the drawing in both horizontal and vertical directions. Some operators prefer cursors hairs to be shorter. The length of the cursor hairs can be adjusted in the Display sub-menu of the Options dialog.

3. Buttons at the left-hand end of the status bar

A number of buttons at the left-hand end of the status bar can be used for toggling (turning on/off) various functions when operating within AutoCAD (Figure 7.1.9). A click on a button turns that function on, if it is off; a click on a button when it is on turns the function off. Similar results can be obtained by using the function keys of the computer keyboard (keys F1 to F10).

Figure 7.1.9 The buttons in left-hand end of the status bar

1) **Snap Mode**: also toggled using the F9 key. When snap is on, the cursor under mouse control can only be moved in jumps from one snap point to another.

2) **Grid Display**: also toggled using the F7 key. When set on, a series of grid points appears in the drawing area.

3) **Ortho Mode**: also toggled using the F8 key. When set on, lines etc. can only be drawn vertically or horizontally.

4) **Polar Tracking**: also toggled using the F10 key. When set on, a small tip appears showing the direction and length of lines etc. in degrees and units.

5) **Object Snap**: also toggled using the F3 key. When set on, an osnap icon appears at the cursor pick box.

6) **Object Snap Tracking**: when set on, lines etc. can be drawn at exact coordinate points and precise angles.

7) **Allow/Disallow Dynamic UCS**: also toggled by the F6 key. Used when constructing 3D solid models.

8) **Dynamic Input**: also toggled by F12. When set on, the x, y coordinates and prompts show when the cursor hairs are moved.

9) **Show/Hide Line weight**: when set on, line weights show on screen. When set off, line weights only show in plotted/printed drawings.

10) **Quick Properties**: a right-click brings

up a pop-up menu, from which a click on Settings causes the Drafting Settings dialog to appear.

Note:

When constructing drawings in AutoCAD, it is advisable to toggle between Snap, Ortho, Osnap and the other functions in order to make constructing easier.

4. The AutoCAD coordinate system

In the AutoCAD 2D coordinate system, units are measured horizontally in terms of X and vertically in terms of Y. A 2D point in the AutoCAD drawing area can be determined in terms of X, Y (in this book referred to as x, y). x, $y = 0$, 0 is the origin of the system. The coordinate point x, $y = 100$, 50 is 100 units to the right of the origin and 50 units above the origin. The point x, $y = -100$, -50 is 100 units to the left of the origin and 50 points below the origin. Figure 7.1.10 shows some 2D coordinate points in the AutoCAD window.

3D coordinates include a third coordinate (Z), in which positive Z units are towards the operator as if coming out of the monitor screen and negative Z units are going away from the operator as if towards the interior of the screen. 3D coordinates are stated in terms of x, y, z. x, y, $z = 100$, 50, 50 is 100 units to the right of the origin, 50 units above the origin and 50 units towards the operator. A 3D model drawing as if resting on the surface of a monitor is shown in Figure 7.1.11.

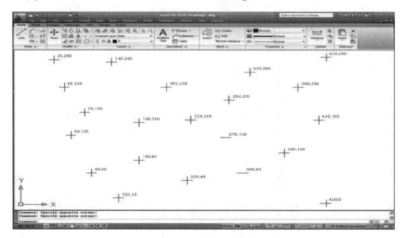

Figure 7.1.10 The 2D coordinate points in the AutoCAD coordinate system

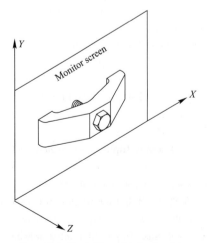

Figure 7.1.11 A 3D model drawing showing the X, Y and Z coordinate directions

5. Tools and tool icons

In AutoCAD, tools are shown as names and icons in panels, in toolbars or in drop-down menus. When the cursor is placed over a tool icon a description shows with the name of the tool as shown and an explanation in diagram form as in the example given in Figure 7.1.6.

If a small outward-facing arrow is included at the right-hand side of a tool icon, when the cursor is placed over the icon and the pick button of the mouse depressed and held, a fly out appears which includes other features. An example is given in Figure 7.1.7.

6. Revision notes

1) A double-click on the AutoCAD short-

cut in the Windows desktop opens the AutoCAD window.

2) There are three main workspaces in which drawings can be constructed——the 2D Drafting & Annotation, Classic AutoCAD and the 3D Modeling workspace. Part 1, 2D Design of this book deals with 2D drawings that will be constructed mainly in the 2D Drafting & Annotation workspace. In Part 2, 3D Design, 3D model drawings will be mainly constructed in the 3D Modeling workspace.

3) All constructions in this book involve the use of a mouse as the digitizer. When a mouse is the digitizer:

A left-click means pressing the left-hand button (the Pick button).

A right-click means pressing the right-hand button (the Return button).

A double-click means pressing the left-hand button twice in quick succession.

Dragging means moving the mouse until the cursor is over an item on screen, holding the left-hand button down and moving the mouse. The item moves in sympathy to the mouse movement.

To pick has a similar meaning to a left-click.

4) Palettes are a particular feature of Auto-CAD. The Command palette and the Design Center palette will be in frequent use.

5) Tools are shown as icons in the tool panels.

6) When a tool is picked, a tooltip appears describing the action of the tool. Most tools show a small tooltip, followed shortly afterwards by a larger one.

7) Dialogs allow opening and saving of Files and the setting of parameters.

8) A number of right-click menus are used in AutoCAD.

9) A number of buttons in the status bar can be used to toggle features such as snap and grid. Functions keys of the keyboard can be also used for toggling some of these functions.

10) The AutoCAD coordinate system determines the position in units of any 2D point in the drawing area (2D Drafting & Annotation and Classic AutoCAD) and any point in 3D space (3D Modeling).

11) Drawings are usually constructed in templates with predetermined settings. Some templates include borders and title blocks.

Note:

Throughout this book when tools are to be selected from panels in the ribbon the tools will be shown in the form e. g. Home/Draw——the name of the tab in the ribbon title bar, followed by the name of the panel from which the tool is to be selected.

7.2 Introducing Drawings

1. The 2D Drafting & Annotation workspace

Illustrations throughout this chapter will be shown as working in the 2D Drafting & Annotation workspace. In this workspace, the Home/Draw panel is at the left-hand end of the Ribbon and Draw tools can be selected from the panel as indicated by a click on the Line tool (Figure 7.2.1). In this chapter all examples will show tools as selected from the Home/Draw panel. However, methods of construction will be the same if the reader wishes to work in other workspaces. When working in the Classic AutoCAD workspace, Draw tools can be selected from the Draw toolbar (Figure 7.2.2). The Draw tools can also be selected from the Draw drop-down menu whether working in 2D Drafting & Annotation or Classic AutoCAD.

2. Drawing with the line tool

1) First example——Line tool (Figure 7.2.3)

(1) Open AutoCAD. The drawing area will show the settings of the acadiso, dwt template–Limits set to 420, 297, Grid set to 10,

Snap set to 5 and Units set to 0.

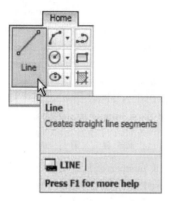

Figure 7.2.1　The Line tool from the Home/Draw panel

(2) Left-click on the Line tool in the Home/Draw panel, in the Draw toolbar (Figure 7.2.1 and Figure 7.2.2) or enter line or l at the command line.

Notes:

① The tooltip appears when the tool icon is clicked.

② The prompt Command: _ line Specify first point appears in the command window at the command line (Figure 7.2.4).

(3) Make sure Snap is on by either pressing the F9 key or the Snap Mode button in the status bar. <Snap on> will show in the command palette.

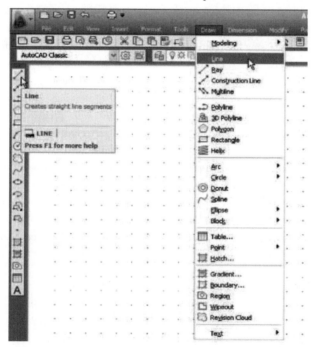

Figure 7.2.2　Selecting the Line tool in the Classic AutoCAD workspace

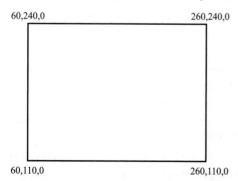

Figure 7.2.3　First example——Line tool

(4) Move the mouse around the drawing area. The cursors pick box will jump from point to point at 5 unit intervals. The position of the pick box will show as coordinate numbers in the status bar (left-hand end).

(5) Move the mouse until the coordinate numbers show 60, 240, 0 and press the pick button of the mouse (left-click).

(6) Move the mouse until the coordinate numbers show 260, 240, 0 and left-click.

(7) Move the mouse until the coordinate

numbers show 260, 110, 0 and left-click.

(8) Move the mouse until the coordinate numbers show 60, 110, 0 and left-click.

(9) Move the mouse until the coordinate numbers show 60, 240, 0 and left- click. Then press the Return button of the mouse (right-click).

The line rectangle Figure 7.2.3 appears in the drawing area.

2) Second example——Line tool (Figure 7.2.5)

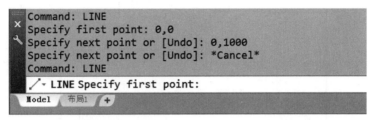

Figure 7.2.4　The prompt appearing at the command line in the Command palette

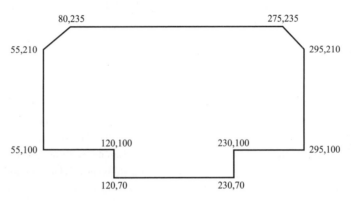

Figure 7.2.5　Second example——Line tool

(1) Clear the drawing from the screen with a click on the drawing Close button of the AutoCAD drawing area. Make sure it is not the AutoCAD window button.

(2) The warning window Figure 7.2.6 appears in the center of the screen. Click its "No" button.

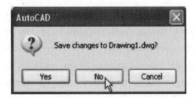

Figure 7.2.6　The AutoCAD warning window

(3) Left-click New...... button in the File drop-down menu and from the Select template dialog which appears double-click on acadiso. dwt.

(4) Left-click on the Line tool icon and enter figures as follows at each prompt of the command line sequence:

Command: line Specify first point: enter 80, 235 right-click.

Specify next point or [Undo]: enter 275, 235 right-click.

Specify next point or [Undo]: enter 295, 210 right-click.

Specify next point or [Close/Undo]: enter 295, 100 right-click.

Specify next point or [Close/Undo]: enter 230, 100 right-click.

Specify next point or [Close/Undo]: enter 230, 70 right-click.

Specify next point or [Close/Undo]: enter 120, 70 right-click.

Specify next point or [Close/Undo]: enter 120, 100 right-click.

Specify next point or [Close/Undo]: enter 55, 100 right-click.

Specify next point or [Close/Undo]: enter

55, 210 right-click.

Specify next point or [Close/Undo]: enter c (Close) right-click.

The result is shown in Figure 7.2.5.

3) Third example——Line tool (Figure 7.2.7)

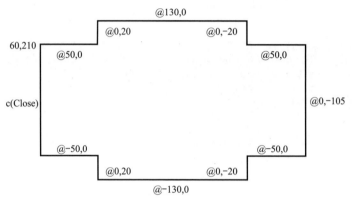

Figure 7.2.7 Third example——Line tool

(1) Close the drawing and open a new acadiso. dwt window.

(2) Left-click on the Line tool icon and enter figure as follows at each prompt of the command line sequence:

Command: _ line Specify first point: enter 60, 210 right-click.

Specify next point or [Undo]: enter @50, 0 right-click.

Specify next point or [Undo]: enter @0, 20 right-click.

Specify next point or [Close/Undo]: enter @130, 0 right-click.

Specify next point or [Close/Undo]: enter @0, −20 right-click.

Specify next point or [Close/Undo]: enter @50, 0 right-click.

Specify next point or [Close/Undo]: enter @0, −105 right-click.

Specify next point or [Close/Undo]: enter @−50, 0 right-click.

Specify next point or [Close/Undo]: enter @0, −20 right-click.

Specify next point or [Close/Undo]: enter @−130, 0 right-click.

Specify next point or [Close/Undo]: enter @0, 20 right-click.

Specify next point or [Close/Undo]: enter @−50, 0 right-click.

Specify next point or [Close/Undo]: enter c (Close) right-click.

The result is as shown in Figure 7.2.7.

Notes:

(1) The figures typed at the keyboard determining the corners of the outlines in the above examples are two-dimensional (2D) x, y coordinate points. When working in 2D, coordinates are expressed in terms of two numbers separated by a comma.

(2) Coordinate points can be shown as positive or as negative numbers.

(3) The method of constructing an outline as shown in the first two examples above is known as the absolute coordinate entry method, where the x, y coordinates of each corner of the outlines are entered at the command line as required.

(4) The method of constructing an outline as shown in the third example is known as the relative coordinate entry method——coordinate points are entered relative to the previous entry. In relative coordinate entry, the @ symbol is entered before each set of coordinates with the following rules in mind:

+ vex entry is to the right.

− vex entry is to the left.

+ vey entry is upwards.

− vey entry is downwards.

(5) The next example (the fourth) shows how lines at angles can be drawn taking advan-

tage of the relative coordinate entry method Angles in AutoCAD are measured in 360 degrees in a counter- clockwise (anticlockwise) direction (Figure 7.2.8). The "<" symbol precedes the angle.

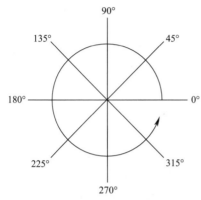

Figure 7.2.8　The counter-clockwise direction of measuring angles in AutoCAD

4) Fourth example——Line tool (Figure 7.2.9)

(1) Close the drawing and open a new acadiso. dwt window.

(2) Left-click on the Line tool icon and enter figures as follows at each prompt of the command line sequence:

Command: _ line Specify first point: enter 70, 230 Specify next point: right-click @220, 0.
Specify next point: right-click @ 0, -70
Specify next point or [Undo]: right-click @ 115 < 225.
Specify next point or [Undo]: right-click @ -60, 0 Specify next point or [Close/Undo]: right-click @ 115 < 135.
Specify next point or [Close/Undo]: right-click @ 0, 70.
Specify next point or [Close/Undo]: right-click c (Close).

The result is as shown in Figure 7.2.9.

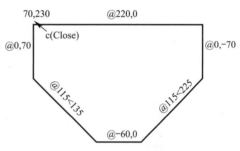

Figure 7.2.9　Fourth example——Line tool

5) Fifth example——Line tool (Figure 7.2.10)

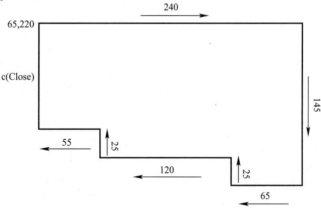

Figure 7.2.10　Fifth example——Line tool

Another method of constructing accurate drawings is by using a method known as tracking. When Line is in use, as each Specify next point: appears at the command line, a rubber-banded line appears from the last point entered. Drag the rubber-band line in any direction and enter a number at the keyboard, followed by a right-click. The line is drawn in the dragged direction of a length in units equal to the entered number.

In this example because all lines are drawn in vertical or horizontal directions, either press the F8 key or click the ORTHO button in the status bar which will only allow drawing horizontally or vertically.

(1) Close the drawing and open a new acadiso. dwt window.

(2) Left-click on the Line tool icon and enter figures as follows at each prompt of the command line sequence:

Command: _ line Specify first point: enter 65, 220 right-click.

Specify next point: drag to right enter 240 right-click.

Specify next point: drag down enter 145 right-click.

Specify next point or [Undo]: drag left enter 65 right-click.

Specify next point or [Undo]: drag upwards enter 25 right-click.

Specify next point or [Close/Undo]: drag left enter 120 right-click.

Specify next point or [Close/Undo]: drag upwards enter 25 right-click.

Specify next point or [Close/Undo]: drag left enter 55 right-click.

Specify next point or [Close/Undo]: c (Close) right-click.

The result is as shown in Figure 7.2.10.

3. Drawing with the circle tool

1) First example——Circle tool (Figure 7.2.13)

(1) Close the drawing just completed and open the acadiso. dwt template.

(2) Left-click on the Circle tool icon in the Home/Draw panel (Figure 7.2.11).

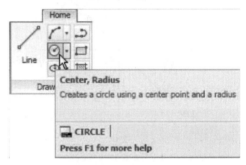

Figure 7.2.11 The Circle tool from the Home/Draw panel

(3) Enter a coordinate and a radius against the prompts appearing in the command window as shown in Figure 7.2.12, followed by right-clicks. The circle (Figure 7.2.13) appears on screen.

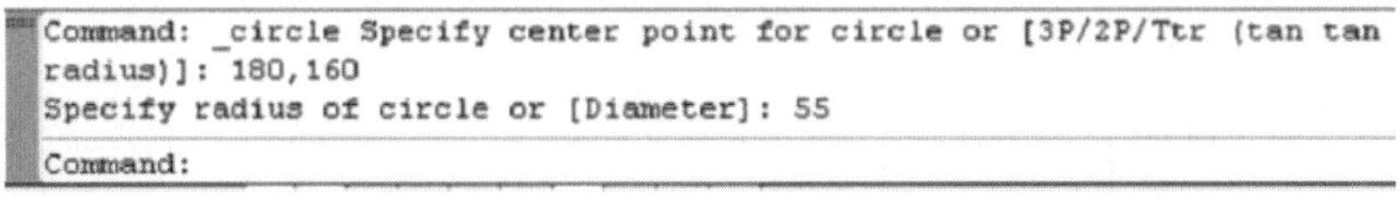

```
Command: _circle Specify center point for circle or [3P/2P/Ttr (tan tan radius)]: 180,160
Specify radius of circle or [Diameter]: 55
Command:
```

Figure 7.2.12 First example——Circle. The command line prompts when Circle is called

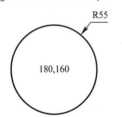

Figure 7.2.13 First example——Circle tool

2) Second example——Circle tool (Figure 7.2.14)

(1) Close the drawing and open the acadiso. dwt screen.

(2) Left-click on the Circle tool icon and construct two circles as shown in the drawing Figure 7.2.14 in the positions and radii shown in Figure 7.2.15.

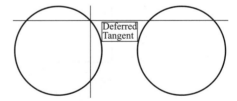

Figure 7.2.14 Second example——Circle tool-the two circles of radius 50

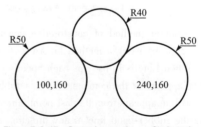

Figure 7.2.15 Second example——Circle tool

(3) Click the Circle tool again and against the first prompt enter t (the abbreviation for the prompt tan tan radius), followed by a right-click.

Command_ circle Specify center point for circle or [3P/2P/Ttr (tan tan radius)]: enter t right-click.

Specify point on object for first tangent of circle: pick.

Specify point on object for second tangent of circle: pick.

Specify radius of circle (50): enter 40 right-click.

The radius 40 circle tangential to the two circle already drawn then appears (Figure 7.2.15).

Notes:

When a point on either circle is picked a tip (Deferred Tangent) appears. This tip will only appear when the Object Snap button is set on with a click on its button in the status bar, or the F3 key of the keyboard is pressed.

Circles can be drawn through 3 points or through 2 points entered at the command line in response to prompts brought to the command line by using 3P and 2P in answer to the circle command line prompts.

4. The Erase tool

If an error has been made when using any of the AutoCAD tools, the object or objects which have been incorrectly drawn can be deleted with the Erase tool. The Erase tool icon can be selected from the Home/ Modify panel (Figure 7.2.16) or by entering e at the command line.

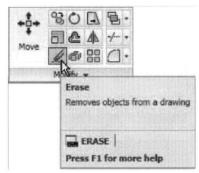

Figure 7.2.16 The Erase tool icon from the Home/Modify panel

1) First example——Erase (Figure 7.2.17)

(1) With Line construct the outline Figure 7.2.18.

(2) Assuming two lines of the outline have been incorrectly drawn, left- click the Erase tool icon. The command line shows:

Command: _ erase.

Select objects: pick one of the lines 1 found Select objects: pick the other line 2 total. Select objects: right-click.

And the two lines are deleted (right-hand drawing of Figure 7.2.17).

2) Second example——Erase (Figure 7.2.19)

The two lines could also have been deleted by the following method:

(1) Left-click the Erase tool icon. The command line shows:

Command: _ erase.

Select objects: enter c (Crossing) Specify first corner: pick.

Specify opposite corner: pick.

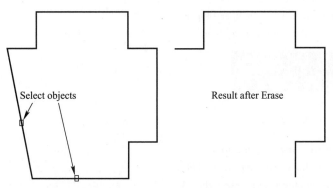

Figure 7.2.17   First example —— Erase

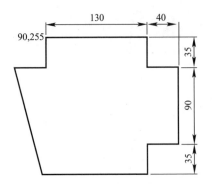

Figure 7.2.18 First example——Erase. An incorrect outline

(2) found Select objects: right-click.

And the two lines are deleted as in the right-hand drawing Figure 7.2.19.

5. Undo and Redo tools

Two other tools of value when errors have been made are the Undo and Redo tools. To undo the last action taken by any tool when constructing a drawing, either left-click the Undo tool in the Quick Access toolbar (Figure 7.2.20) or enter u at the command line. No matter which method is adopted the error is deleted from the drawing.

Everything constructed during a session of drawing can be undone by repeated clicking on the Undo tool icon or by repeatedly entering u's at the command line.

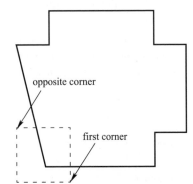

Figure 7.2.19 Second example——Eras

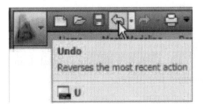

Figure 7.2.20 The Undo tool in the Title bar

To bring back objects that have just been removed by the use of Undo's, left-click the Redo tool icon in the Quick Access toolbar (Figure 7.2.21) or enter redo at the command line.

Figure 7.2.21 The Redo tool icon in the Title bar

6. Drawing with the Polyline tool

When drawing lines with the Line tool, each line drawn is an object in its own right. A rectangle drawn with the Line tool is four objects. A rectangle drawn with the Polyline tool is a single object. Lines of different thickness, arcs, arrows and circles can all be drawn using this tool, as will be shown in the examples describing constructions using the Polyline tool. Constructions resulting from using the tool are known as polylines or plines.

The Polyline tool can be called from the Home/Draw panel (Figure 7.2.22) or by entering pl at the command line.

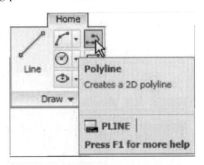

Figure 7.2.22 The Polyline tool icon in the Home/Draw panel

1) First example——Polyline tool (Figure 7.2.23)

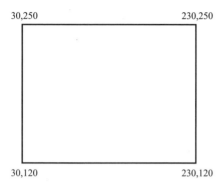

Figure 7.2.23 First example —— Polyline tool

Notes:

In this example enter and right-click have not been included. Left-click the Polyline tool. The command line shows:

Command: _ pline Specify start point: 30, 250 Current line width is 0.

Specify next point or [Arc/Half width/ Length/ Undo/Width]: 230, 250.

Specify next point or [Arc/Close/Half width/ Length/Undo/Width]: 230, 120.

Specify next point or [Arc/Close/Half width/ Length/Undo/Width]: 30, 120.

Specify next point or [Arc/Close/Half width/ Length/Undo/Width]: c (Close).

Notes:

(1) Note the prompts: Arc for constructing pline arcs; Close to close an outline; Half width to halve the width of a wide pline; Length to enter the required length of a pline; Undo to undo the last pline constructed; Close to close an outline.

(2) Only the capital letter(s) of a prompt needs to be entered in upper or lower case to make that prompt effective.

(3) Other prompts will appear when the Polyline tool is in use as will be shown in later examples.

2) Second example——Polyline tool (Figure 7.2.24)

This will be a long sequence, but it is typical of a reasonably complex drawing using the Polyline tool. In the following sequences, when a prompt line is to be repeated, the prompts in square brackets ([ ]) will be replaced by [prompts].

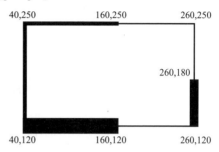

Figure 7.2.24 Second example——Polyline tool

Left-click the Polyline tool icon. The command line shows:

Command: _ pline Specify start point: 40, 250 Current line width is 0.

Specify next point or [Arc/Half width/ Length/Undo/ Width]: w (Width).

Specify starting width <0>: 5 Specify ending width <5>: right-click.

Specify next point or [Arc/Close/Half width/Length/Undo/ Width]: 160, 250.

Specify next point or [prompts]: h (Half width) Specify starting half-width <2.5>: 1.

Specify ending half-width <l>: right-click Specify next point or [prompts]: 260, 250.

Specify next point or [prompts]: 260, 180 Specify next point or [prompts]: w (Width) Specify starting width <l>: 10.

Specify ending width <10>: right-click Specify next point or [prompts]: 260, 120.

Specify next point or [prompts]: h (Half width) Specify starting half-width <5>: 2.

Specify ending half-width <2>: right-click Specify next point or [prompts]: 160, 120.

Specify next point or [prompts]: w (Width) Specify starting width <4>: 20.

Specify ending width <20>: right-click Specify next point or [prompts]: 40, 120 Specify starting width <20>: 5 Specify ending width <5>: right-click.

Specify next point or [prompts]: c (Close) Command.

3) Third example——Polyline tool (Figure 7.2.25)

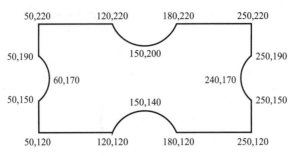

Figure 7.2.25 Third example——Polyline tool

Left-click the Polyline tool icon. The command line shows:

Command: _ pline Specify start point: 50, 220 Current line width is 0 [prompts]: w (Width).

Specify starting width <0>: 0.5 Specify ending width <0.5>: right-click Specify next point or [prompts]: 120, 220 Specify next point or [prompts]: a (Arc).

Specify endpoint of arc or [prompts]: s (second pt) Specify second point on arc: 150, 200.

Specify end point of arc: 180, 220.

Specify end point of arc or [prompts]: l (Line) Specify next point or [prompts]: 250, 220 Specify next point or [prompts]: 260, 190 Specify next point or [prompts]: a (Arc).

Specify endpoint of arc or [prompts]: s (second pt) Specify second point on arc: 240, 170.

Specify end point of arc: 250, 160.

Specify end point of arc or [prompts]: l (Line) Specify next point or [prompts]: 250, 150 Specify next point or [prompts]: 250, 120.

And so on until the outline Figure 2.25 is completed.

4) Fourth example——Polyline tool (Figure 7.2.26)

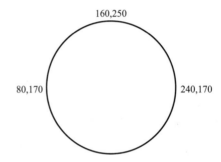

Figure 7.2.26 Fourth example——Polyline tool

Left-click the Polyline tool icon. The command line shows:

Command: _ pline Specify start point: 80, 170 Current line width is 0.

Specify next point or [prompts]: w (Width) Specify starting width <0>: l

Specify ending width <l>: right-click.

Specify next point or [prompts]: a (Arc).

Specify endpoint of arc or [prompts]: s (second pt) Specify second point on arc: 160, 250.

Specify end point of arc: 240, 170.

Specify end point of arc or [prompts]: c (Close).

And the circle Figure 7.2.26 is formed.

5) Fifth example——Polyline tool (Figure 7.2.27)

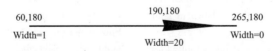

Figure 7.2.27 Fifth example——Polyline tool

Left-click the Polyline tool icon. The command line shows:

Command: _ pline Specify start point: 60, 180 Current line width is 0.

Specify next point or [prompts]: *w* (Width) Specify starting width <0>: *l*.

Specify ending width < *l* >: right-click Specify next point or [prompts]: 190, 180. Specify next point or [prompts]: *w* (Width) Specify starting width <*l*>: 20.

Specify ending width <20>: 0.

Specify next point or [prompts]: 265, 180 Specify next point or [prompts]: right-click.

And the arrow Figure 7.2.27 is formed.

7. Revision notes

1) The following terms have been used in this chapter

Left-click——press the left-hand button of the mouse.

Click——same meaning as left-click.

Double-click——press the left-hand button of the mouse twice.

Right-click——press the left-hand button of the mouse——has the same result as pressing the Return key of the keyboard.

Drag——move the cursor on to an object and, holding down the right-hand button of the mouse, pull the object to a new position.

Enter——type the letters or numbers which follow at the keyboard.

Pick——move the cursor on to an item on screen and press the left-hand button of the mouse.

Return——press the Enter key of the keyboard. This key may also be marked with a left facing arrow. In most cases (but not always) has the same result as a right-click.

Dialog——a window appearing in the AutoCAD window in which settings may be made.

Drop-down menu——a menu appearing when one of the names in the menu bars is clicked.

Tooltip——the name of a tool appearing when the cursor is placed over a tool icon from a toolbar.

Prompts——text appearing in the command window when a tool is selected which advise the operator as to which operation is required.

2) Three methods of coordinate entry have been used in this chapter.

Absolute method——the coordinates of points on an outline are entered at the command line in response to prompts.

Relative method——the distances in coordinate units are entered preceded by @ from the last point which has been determined on an outline. Angles, which are measured in a counter-clockwise direction, are preceded by "<".

Tracking——the rubber band of the tool is dragged in the direction in which the line is to be drawn and its distance in units is entered at the command line followed by a right-click.

Line and Polyline tools——an outline drawn using the Line tool consists of a number of objects——the number of lines in the outline. An outline drawn using the Polyline is a single object.

7.3 Draw tools, Object Snap and Dynamic Input

1. Introduction

The aims of this chapter are:

1) To give examples of the use of the Arc, Ellipse, Polygon, Rectangle tools from the Home/Draw panel.

2) To give examples of the uses of the Polyline Edit (pedit) tool.

3) To introduce the Object Snaps (osnap) and their uses.

4) To introduce the Dynamic Input (DYN) system and its uses.

The majority of tools in AutoCAD can be called into use by any one of the following five methods:

1) By clicking on the tool's icon in the appropriate panel. Figure 7.3.1 shows the Polygon tool called from the Home/Draw panel.

2) When working in the Classic AutoCAD

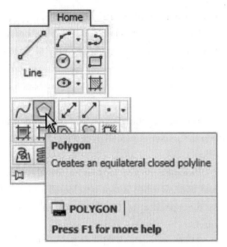

Figure 7.3.1 The Polygon tool and its tooltip selected from the Home/Draw Panel

toolbar. Figure 7.3.2 shows the Draw toolbar. Placing the cursor on the Polygon tool icon in this toolbar shows the same tooltip as that shown in Figure 7.3.2.

3) By clicking on the tool's name in an appropriate drop-down menu. Figure 7.3.3 shows the tool names and icons displayed in the Draw drop-down menu. It is necessary to first bring the menu bar to screen with a click on Show Menu Bar in the left-click menu of the Quick Access Toolbar (Figure 7.3.4) if the menu bar is not already on screen.

4) By entering an abbreviation for the tool name at the command line in the command palette. For example the abbreviation for the Line tool is l, for the Polyline tool it is pl and for the Circle tool it is c.

screen, with a click on the tool's name in a

Figure 7.3.2 The tool icons in the Draw toolbar

5) By entering the full name of the tool at the command line.

6) By making use of the Dynamic Input method of constructing drawings.

Figure 7.3.3 The Draw drop-down menu left-click

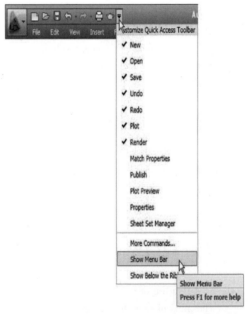

Figure 7.3.4 Selecting Show Menu Bar from the menu in the Quick Access Toolbar

In practice operators constructing drawings in AutoCAD may well use a combination of these six methods.

2. The Arc tool

In AutoCAD, arcs can be constructed using

any three of the following characteristics of an arc: its Start point; a point on the arc (Second point); its Center; its End; its Radius; the Length of the arc; the Direction in which the arc is to be constructed; the Angle between lines of the arc.

These characteristics are shown in the menu appearing with a click on the Arc tool in the Home/Draw panel (Figure 7.3.5).

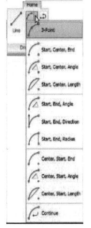

Figure 7.3.5 The Arc tool fly out in the Home/ Draw panel

To call the Arc tool click on the fly out of its tool icon in the Home/Draw panel, click on Arc in the Draw toolbar, click on Arc in the Draw drop-down menu, or enter a or arc at the command line. In the following examples initials of command prompts will be shown instead of selections from the menu shown in Figure 7.3.6.

1) First example——Arc tool (Figure 7.3.6)

Left-click the Arc tool icon. The command line shows:

Command: _ arc Specify start point of arc or [Center]: 100, 220.

Specify second point of arc or [Center/End]: 55, 250 Specify end point of arc: 10, 220.

2) Second example——Arc tool (Figure 7.3.6)

Command: right-click brings back the Arc sequence ARC Specify start point of arc or [Center]: c (Center).

Specify center point of arc: 200, 190 Specify start point of arc: 260, 215.

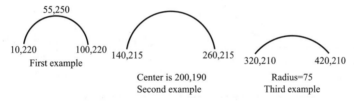

Figure 7.3.6 Examples——Arc tool

Specify end point of arc or [Angle/chord Length]: 140, 215.

3) Third example——Arc tool (Figure 7.3.6)

Command: right-click brings back the Arc sequence ARC Specify start point of arc or [Center]: 420, 210.

Specify second point of arc or [Center/End]: e (End).

Specify end point of arc: 320, 210 Specify center point of arc or.

[Angle/Direction/Radius]: r (Radius) Specify radius of arc: 75.

3. Snap

In previous chapters several methods of constructing accurate drawings have been described——using Snap; absolute coordinate entry; relative coordinate entry and tracking. Other methods of ensuring accuracy between parts of constructions are by making use of Object Snaps (Osnaps).

Snap Mode, Grid Display and Object Snaps can be toggled on/off from the buttons in the status bar or by pressing the keys F9 (Snap Mode), F7 (Grid Display) and F3 (Object Snap), F5 Object Snaps.

Object Snaps allow objects to be added to a drawing at precise positions in relation to other objects already on screen. With Object Snaps, objects can be added to the end points, mid points, to intersections of objects, to centers and quadrants of circles and so on. Object Snaps also override snap points even when snap is set on.

To set Object Snaps–at the command line:
Command: enter os.

And the Drafting Settings dialog appears (Figure 7.3.7). Click the Object Snap tab in the upper part of the dialog and click the check boxes to the right of the Object Snap names to set them on (or off in on).

When Object Snaps are set ON, as outlines are constructed using Object Snaps so Object Snap icons and their tooltips appear as indicated in Figure 7.3.8.

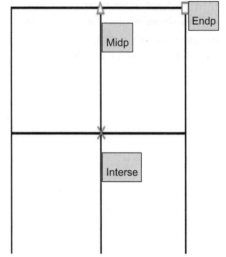

Figure 7.3.8 Three Object Snap icons and their tooltips

1) First example——Object Snap (Figure 7.3.9)

Call the Polyline tool:
Command: _ pline.
Specify start point: 50, 230 [prompts]:
w (Width) Specify starting width: l.
Specify ending width < l >: right-click Specify next point: 260, 230.

Figure 7.3.7 The Drafting Settings dialog with some of the Object Snaps set on

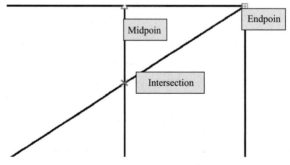

Figure 7.3.9 First example——Osnaps

Specify next point: right-click Command: right-click PLINE.

Specify start point: pick the right-hand end of the pline.

Specify next point: 50, 120 Specify next point: right-click Command: right-click PLINE.

Specify start point: pick near the middle of first line.

Specify next point: 155, 120 Specify next point: right-click Command: right-click PLINE.

Specify start point: pick the plines at their intersection.

Specify start point: right-click.

The result is shown in Figure 7.3.9. In this illustration the Object Snap tooltips are shown as they appear when each object is added to the outline.

It is sometimes advisable not to have Object Snaps set on in the Drafting Settings dialog, but to set Object Snap off and use Object Snap abbreviations at the command line when using tools. The following example shows the use of some of these abbreviations.

2) Second example——Object Snap abbreviations (Figure 7.3.10)

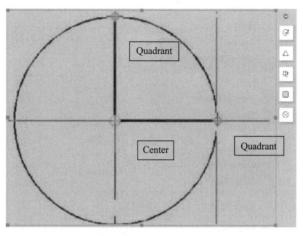

Figure 7.3.10 Second example——Osnaps

Call the Circle tool:
Command: _ circle.
Specify center point for circle: 180, 170.
Specify radius of circle: 60 Command: enter l (Line) right-click.
Specify first point: enter qua right-click of pick near the upper quadrant of the circle Specify next point: enter cen right-click of pick near the centre of the circle.
Specify next point: enter qua right-click of pick near right-hand side of circle Specify next point: right-click.

Note:

With Object Snaps off, the following abbreviations can be used:
end——endpoint;
mid——midpoint;
int——intersection;
cen——centre;
qua——quadrant;
nea——nearest;
ext——extension.

7.4 Zoom, Pan and Templates

1. Introduction

The aims of this chapter are:
1) To demonstrate the value of the Zoom tools.
2) To introduce the Pan tool.
3) To describe the value of using the Aerial View window in conjunction with the Zoom and Pan tools.
4) To update the acadiso. dwt template.
5) To describe the construction and saving of drawing templates.

The use of the Zoom tools allows not only the close inspection of the most minute areas of a drawing in the AutoCAD drawing area, but allows the accurate construction of very small details in a drawing.

The Zoom tools can be called by clicking the Zoom button in the status bar (Figure 7.4.1) or selecting from the View/Zoom drop down menu. However by far the easiest and quickest method of calling the Zoom is to enter z at the command line as follows:

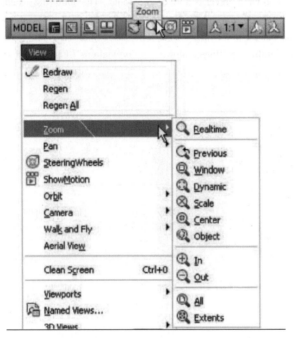

Figure 7.4.1 Calling Zoom——by clicking the Zoom button in the status bar or from the View drop-down menu

Command: enter z right-click.

Zoom Specify corner of window, enter a scale factor (nX or nXP) or [All/ Center/ Dynamic/Extents/Previous/ Scale/Window/ Object] < real time>.

This allows the different zooms:

Realtime——selects parts of a drawing within a window.

All——the screen reverts to the limits of the template.

Center——the drawing centers itself around a picked point.

Dynamic——a broken line surrounds the drawing which can be changed in size and repositioned to another part of the drawing.

Extents——the drawing fills AutoCAD drawing area.

Previous——the screen reverts to its previous zoom.

Scale——entering a number or a decimal fraction scales the drawing.

Window——the parts of the drawing within a picked window appears on screen. The effect is the same as using real time.

Object——pick any object on screen and the object zooms.

The operator will probably be using Realtime, Window and Previous zooms most frequently.

Figure 7.4.2–Figure 7.4.4 show a drawing which has been constructed, a Zoom Window of part of the drawing allowing it to be checked for accuracy, and a Zoom Extents respectively.

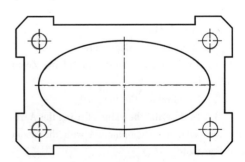

Figure 7.4.2 Drawing to be acted upon by the Zoom tool

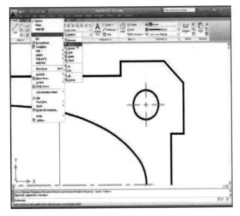

Figure 7.4.3 A Zoom Window of part of the drawing Figure 7.4.2

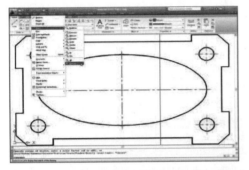

Figure 7.4.4 A Zoom Extents of the drawing Figure 7.4.2

It will be found that the Zoom tools are among those most frequently used when working in AutoCAD.

2. The Pan tool

The Pan tools can be called with a click on the Pan button in the status bar, from the Pan sub-menu of the View drop-down menu or by entering p at the command line. When the tool is called, the cursor on screen changes to an icon of a hand. Dragging the hand across screen under mouse movement allows various parts of the drawing not in the AutoCAD drawing area to be viewed. As the dragging takes place, the green rectangle in the Aerial View window moves in sympathy (see Figure 7.4.5). The Pan tool allows any part of the drawing to be viewed and/or modified. When that part of the drawing which is required is on screen a right-click calls up the menu as shown in Figure 7.4.5, from which either the tool can be exited, or other tools can be called.

Notes:

1) If using a mouse with a wheel both zooms and pans can be performed with the aid of the wheel.

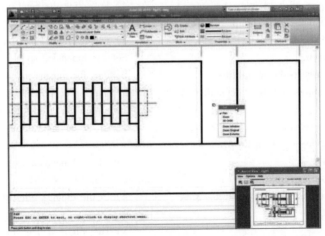

Figure 7.4.5 The Pan tool in action showing a part of the drawing, while the whole drawing is shown in the Aerial View window

2) The Zoom tools are important in that they allow even the smallest parts of drawings to be examined and, if necessary, amended or modified.

3) The Zoom tools can be called with a click on the Zoom button in the status bar, from the Zoom toolbar, from the sub-menu of the View drop-down menu or by entering zoom or z at the command line. The easiest method is to enter z at the command line followed by a right-click.

4) Similarly the easiest method of calling the Pan tool is to enter p at the command line followed by a right-click.

5) When constructing large drawings, the Pan tool and the Aerial View window are of value for allowing work to be carried out in any part of a drawing, while showing the whole drawing in the Aerial View window.

3. Drawing templates

In Section 7.1 to Section 7.3, drawings were constructed in the template acadiso. dwt which loaded when AutoCAD was opened. The default acadiso template has been amended to Limits set to 420, 297 (coordinates within which a drawing can be constructed), Grid Display set to 10, Snap Mode set to 5, and the drawing area Zoomed to All.

Throughout this book most drawings will be based on an A3 sheet, which measures 420 units by 297 units (the same as Limits).

Note:

As mentioned on page before, if others are using the computer on which drawings are being constructed, it is as well to save the template being used to another file name or, if thought necessary to a floppy disk or other temporary type of disk. A file name My_ template. dwt, as suggested earlier, or a name such as book_ template can be given.

Adding features to the template:

Four other features will now be added to our template:

Text style——set in the Text Style dialog.

Dimension style——set in the Dimension Style Manager dialog.

Shortcut menu variable——set to 0.

Layers——set in the Layer Properties Manager dialog.

Setting text:

1) At the command line:

Command: enter st (Style) right-click.

2) The Text style dialog appears (Figure 7.4.6). In the dialog, enter 6 in the Height field. Then left-click on Arial in the Font name pop-up list. Arial font letters appear in the Preview area of the dialog.

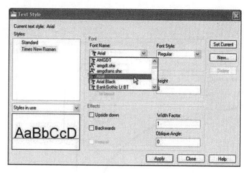

Figure 7.4.6 The Text Style dialog

3) Left-click the New button and enter Arial in the New text style sub-dialog which appears (Figure 7.4.7) and click the OK button.

Figure 7.4.7 The New Text Style sub-dialog

4) Left-click the Set Current button of the Text Style dialog.

5) Left-click the Close button of the dialog.

Setting dimension style:

Settings for dimensions require making entries in a number of sub-dialogs in the Dimension. Style Manager. To set the dimensions style:

1) At the command line: Command: enter d right-click.

And the Dimension Style Manager dialog appears (Figure 7.4.8).

2) In the dialog, click the Modify…… button.

3) The Modify Dimension Style dialog appears (Figure 7.4.9). A number of tabs show at the top of the dialog. Click the Lines tab and make settings as shown in Figure 7.4.9. Then click the OK button of that dialog.

4) The original Dimension Style Manager reappears. Click its Modify button again.

5) The Modify Dimension Style dialog re-

appears (Figure 7.4.10). Click the Symbols and Arrows tab. Set Arrow size to 7.

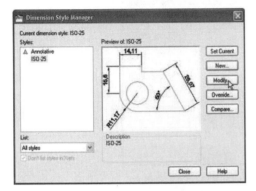

Figure 7.4.8 The Dimension Style Manager dialog

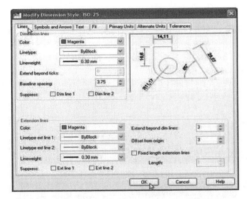

Figure 7.4.9 The setting for Lines in the Modify Dimension Style dialog

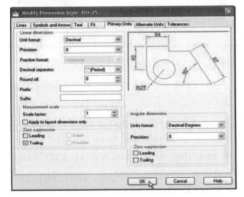

Figure 7.4.10 Setting Primary Units in the Dimension Style Manager

6) Then click the Text tab. Set Text style to Arial, set Color to Magenta, set Text Height to 6 and click the ISO check box in the bottom right-hand corner of the dialog.

7) Then click the Primary Units tab and set the units Precision to 0, that is no units after decimal point and Decimal separator to Period. Click the sub-dialogs OK button (Figure 7.4.10).

8) The Dimension Styles Manager dialog reappears showing dimensions, as they will appear in a drawing, in the Preview of my-style box. Click the New...... button. The Create New Dimension Style dialog appears (Figure 7.4.11).

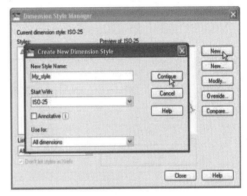

Figure 7.4.11 The Create New Dimension Style dialog

9) Enter a suitable name in the New style name field——in this example this is My-style. Click the Continue button and the Dimension Style Manager appears (Figure 7.4.12). This dialog now shows a preview of the My-style dimensions. Click the dialog's Set Current button, followed by another click on the Close button (Figure 7.4.12).

Setting the shortcut menu variable:

Call the line tool, draw a few lines and then right-click. The right-click menu shown in Figure 7.4.13 may well appear. The menu will also appear when any tool is called. Some operators prefer using this menu when constructing drawings. To stop this menu appearing:

Command: enter shortcut menu right-click.

Enter new value for SHORTCUTMENU < 12>: 0.

And the menu will no longer appear when a tool is in action.

Setting layers:

1) At the command line enter layer followed by a right-click. The Layer Properties Manager palette appears (Figure 7.4.14).

2) Click the New Layer icon. Layer1 ap-

pears in the layer list. Overwrite the name Layer1 entering Centre.

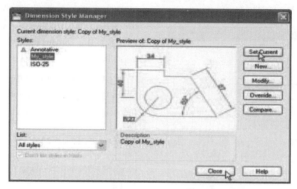

Figure 7.4.12   The Dimension Style Manager reappears. Click the Set Current and Close buttons

Figure 7.4.13   The right-click menu

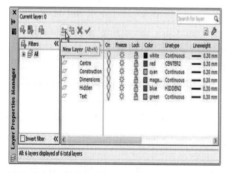

Figure 7.4.14   The Layer Properties Manager palette

3) Repeat step 2 four times and make four more layers entitled.

Construction, Dimensions, Hidden and Text

4) Click one of the squares under the Color column of the dialog. The Select Color dialog appears (Figure 7.4.15). Double-click on one of the colours in the Index Color squares. The selected colour appears against the layer name in which the square was selected. Repeat until all 5 new layers have a colour.

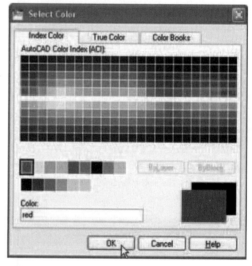

Figure 7.4.15   The Select Color dialog

5) Click on the linetype Continuous against the layer name Centre. The Select Linetype dialog appears (Figure 7.4.16). Click its Load...... button and from the Load or Reload Linetypes dialog double-click CENTER2. The dialog disappears and the name appears in the Select Linetype dialog. Click the OK button and the linetype CENTER2 appears against the layer Centre.

6) Repeat with layer Hidden, load the linetype HIDDEN2 and make the linetype against this layer HIDDEN2.

7) Click on the any of the lineweights in the Layer Properties Manager. This brings up

the Lineweight dialog (Figure 7.4.17). Select the lineweight 0.3. Repeat the same for all the other layers. Then click the Close button of the Layer Properties Manager.

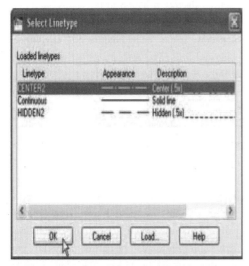

Figure 7.4.16 The Select Linetype dialog

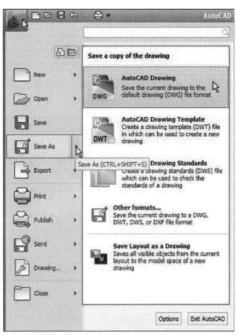

Figure 7.4.18 Calling Save As

2) In the Save Drawing As dialog which comes on screen (Figure 7.4.19), click the arrow to the right of the Files of type field and in the pop-up list associated with the field click on AutoCAD Drawing Template (*.dwt). The list of template files in the AutoCAD/Template directory appears in the file list.

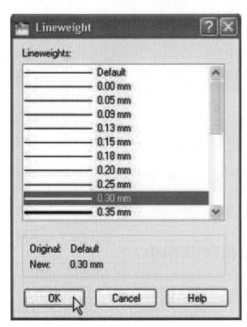

Figure 7.4.17 The Line weight dialog

Saving the template file:

1) Left-click on Save As in the menu appearing with a left-click on the AutoCAD icon at the top left-hand corner of the screen (Figure 7.4.18).

Figure 7.4.19 Saving the template to the name acadiso.dwt

3) Click on acadiso in the file list, followed by a click on the Save button.

4) The Template Option dialog appears. Make entries as suggested in Figure 7.4.20, making sure that Metric is chosen from the pop-up list.

Figure 7.4.20 The Template Description dialog

The template now saved can be opened for the construction of drawings as needed. Now when AutoCAD is opened again the template acadiso. dwt appears on screen.

Note:

Please remember that if others are using the computer, it is advisable to save the template to a name of your own choice.

4. Revison notes

1) The Zoom tools are important in that they allow even the smallest parts of drawings to be examined and, if necessary, amended or modified.

2) The Zoom tools can be called with a click on the Zoom button in the status bar, from the Zoom toolbar, from the sub-menu of the View drop-down menu, or by entering z or zoom at the command line. The easiest method is to enter z at the command line followed by a right-click.

3) There are five methods of calling tools for use——selecting a tool icon in a panel from a group of panels shown as tabs in the Ribbon; selecting a tool from a toolbar; entering the name of a tool in full at the command line; entering an abbreviation for a tool at the command line; selecting a tool from a drop-down menu.

4) When constructing large drawings, the Pan tool and the Aerial View window are of value for allowing work to be carried out in any part of a drawing, while showing the whole drawing in the Aerial View window.

5) An A3 sheet of paper is 420mm by 297mm. If a drawing constructed in the template acadiso. dwt described in this book is printed/plotted full size (scale 1 : 1), each unit in the drawing will be 1mm in the print/plot.

6) When limits are set it is essential to call Zoom followed by a (All) to ensure that the limits of the drawing area are as set.

7) If the right-click menu appears when using tools, the menu can be aborted if required by setting the SHORTCUTMENU variable to 0.

7.5 The Modify Tools

1. Introduction

The aim of this section is to describe the uses of tools for modifying parts of drawings.

The Modify tools are among those most frequently used. The tools are found in the Home/Modify panel. A click on the arrow in the Home/Modify panel brings down a further set of tool icons (Figure 7.5.1). They can also be selected from the Modify toolbar (Figure 7.5.2) or from the Modify drop-down menu.

Using the Erase tool from Home/Modify was described in Section 7.2. Examples of tools other than the Explode follow.

2. The Copy tool

1) First example——Copy

(1) Construct Figure 7.5.3 using Polyline. Do not include the dimensions.

(2) Call the Copy tool——either left-click

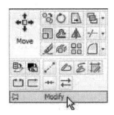

Figure 7.5.1 The Modify tool icons in the Home/Modify panel

Figure 7.5.2 The Modify toolbar

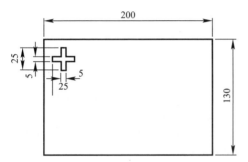

Figure 7.5.3 First example——Copy Object-outlines

on its tool icon in the Home/ Modify panel (Figure 7.5.4), pick Copy from the Modify toolbar, or enter cp or copy at the command line.

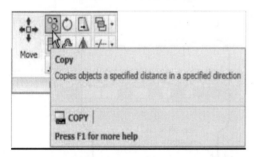

Figure 7.5.4 The Copy tool from the Home/Modify panel

The command line shows:

Command: _ copy.

Select objects: pick the cross 1 found Select objects: right-click.

Current settings: Copy mode = Multiple Specify base point or [Displacement/Mode] <Displacement>: pick.

Specify second point or [Exit/Undo]: pick.

Specify second point or [Exit/Undo] <Exit>: right-click.

The result is given in Figure 7.5.5.

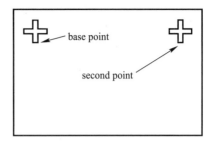

Figure 7.5.5 First example——Copy

2) Second example——Multiple copy (Figure 7.5.6)

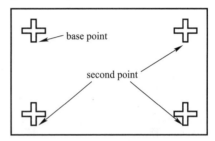

Figure 7.5.6 Second example——Copy-Multiple copy

(1) Erase the copied object.

(2) Call the Copy tool. The command line shows:

Command: _ copy.

Select objects: pick the cross 1 found Select objects: right-click.

Current settings: Copy mode = Multiple Specify base point or [Displacement/Mode] <Displacement>: pick.

Specify second point or <use first point as displacement>: pick.

Specify second point or [Exit/Undo] <Exit>: pick Specify second point or [Exit/Undo] <Exit>: pick Specify second point or [Exit/Undo] <Exit>: e.

The result is shown in Figure 7.5.7.

3. The Mirror tool

1) First example——Mirror

(1) Construct the outline Figure 7.5.7 using the Line and Arc tools.

(2) Call the Mirror tool——left-click on its tool icon in the Home/Modify panel (Fig-

ure 7.5.8), pick Mirror from the Modify toolbar or from the Modify drop-down menu, or enter mi or mirror at the command line. The command line shows:

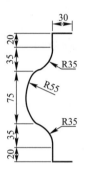

Figure 7.5.7　First example——Mirror-outline

Figure 7.5.8　The Mirror tool from the Modify toolbar

Command: _ mirror.

Select objects: pick first corner Specify opposite corner: pick 7 found.

Select objects: right-click.

Specify first point of mirror line: pick Specify second point of mirror line: pick.

Erase source objects [Yes/No] <N>: right-click.

The result is shown in Figure 7.5.9.

2) Second example——Mirror

(1) Construct the outline shown in the dimensioned polyline in the upper drawing of Figure 7.5.10.

(2) Call Mirror and using the tool three times complete the given outline. The two points shown in Figure 7.5.10 are to mirror the right-hand side of the outline.

3) Third example——Mirror

If text is involved when using the Mirror tool, the set variable MIRRTEXT must be set correctly. To set the variable:

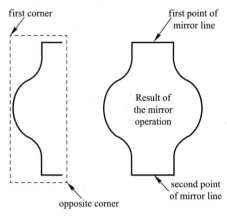

Figure 7.5.9　First example——Mirror

Command: mirrtext.

Enter new value for MIRRTEXT <1>: 0.

If set to 0 text will mirror without distortion. If set to 1 text will read backwards as indicated in Figure 7.5.11.

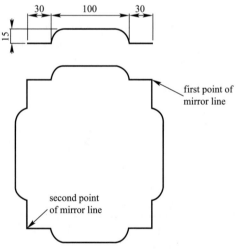

Figure 7.5.10　Second example——Mirror

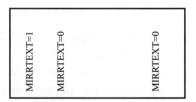

Figure 7.5.11　Third example——Mirror

4. The Offset tool

Examples——Offset:

1) Construct the four outlines shown in Figure 7.5.12.

2) Call the Offset tool——left-click on its tool icon in the Home/Modify panel (Figure 7.5.13), pick the tool from the Modify toolbar, pick the tool name in the Modify drop-down menu, or enter o or offset at the command line. The command line shows:

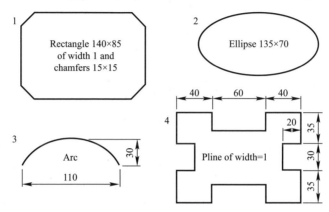

Figure 7.5.12 Examples——Offset-outlines

Command: _ offset.

Current settings: Erase source = No Layer = Source OFFSETGAPTYPE = 0.

Specify offset distance or [Through/Erase/Layer] <Through>: 10.

Select object to offset or [Exit/Undo]<Exit>: pick drawing 1.

Specify point on side to offset or [Exit/Multiple/Undo] <Exit>: pick inside the rectangle.

Select object to offset or [Exit/Undo] <Exit>: e (Exit).

3) Repeat for drawings 2, 3 and 4 in Figure 7.5.12 as shown in Figure 7.5.14.

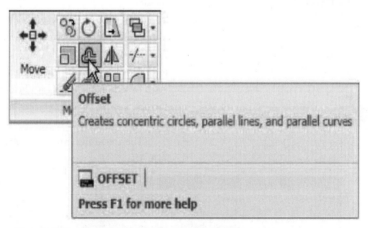

Figure 7.5.13 The Offset tool from the Home/Modify panel

5. The Array tool

Arrays can be in either a Rectangular form or in a Polar form as shown in the examples below.

1) First example——Rectangular Array

(1) Construct the drawing Figure 7.5.15.

(2) Call the Array tool——either click Array in the Modify drop-down menu (Figure 7.5.16), from the Home/Modify panel, pick the Array tool icon from the Modify toolbar, or enter ar or array at the command line. The Array dialog appears (Figure 7.5.17).

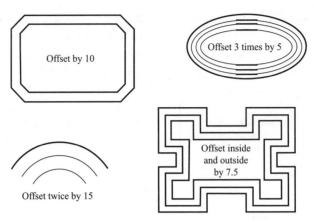

Figure 7.5.14 Examples——Offset

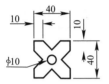

Figure 7.5.15 First example——Array-drawing to be arrayed

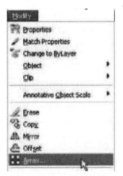

Figure 7.5.16 First example——Call the Array tool

(3) Make settings in the dialog:

Rectangular Array radio button set on (dot in button).

Row field——enter **5**.

Column field——enter **6**.

Row offset field——enter **−50** (note the minus sign).

Column offset field——enter **50**.

(4) Click the Select objects button and the dialog disappears. Window the drawing. A second dialog appears which includes a Preview< button.

(5) Click the Preview< button. The dialog disappears and the following prompt appears at the command line: Pick or press Esc to return to drawing or <Right- click to accept drawing>.

(6) If satisfied right-click. If not, press the Esc key and make revisions to the Array dialog fields as necessary.

The resulting array is shown in Figure 7.5.18.

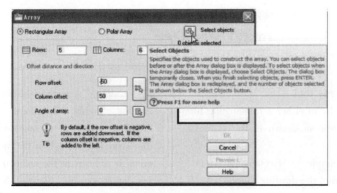

Figure 7.5.17 First example——the Array dialog

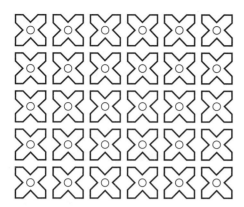

Figure 7.5.18 First example——Array

2) Second example——Polar Array

(1) Construct the drawing Figure 7.5.19.

(2) Call Array. The Array dialog appears. Make settings as shown in Figure 7.5.20.

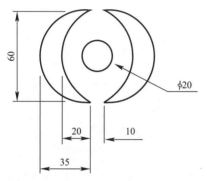

Figure 7.5.19 Second example——the drawing to be arrayed

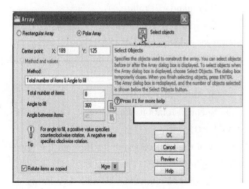

Figure 7.5.20 Second example——Array-settings in the dialog

(3) Click the Select objects button of the dialog and window the drawing. The dialog returns to screen. Click the Pick center point button (Figure 7.5.21) and when the dialog disappears, pick a centre point for the array.

(4) The dialog reappears. Click its Preview< button. The array appears and the command line shows.

Pick or press Esc to return to drawing or < Right- click to accept drawing>.

(5) If satisfied right-click. If not, press the Esc key and make revisions to the Array dialog fields as necessary.

The resulting array is shown in Figure 7.5.22.

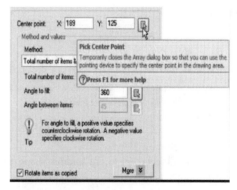

Figure 7.5.21 Second example——Array——the Pick Center point button

Figure 7.5.22 Second example——Array

6. The Move tool

Example——Move:

1) Construct the drawing Figure 7.5.23.

2) Call Move——either click the Move tool icon in the Home/Modify panel (Figure 7.5.24), pick Move from the Modify toolbar, pick Move from the Modify drop-down menu, or

enter m or move at the command line, which shows:

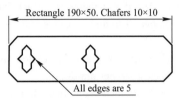

Figure 7.5.23　Example——Move-drawing

Figure 7.5.24　The Move tool from the Home/Modify toolbar

Command: _ move.

Select objects: pick the middle shape in the drawing 1 found.

Select objects: right-click.

Specify base point or [Displacement] <Displacement>: pick.

Specify second point or <use first point as displacement>: pick.

The result is given in Figure 7.5.25.

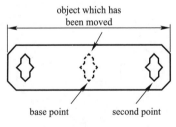

Figure 7.5.25　Example——Move

7. The Rotate tool

When using the Rotate tool remember the default rotation of objects within AutoCAD is counterclockwise (anticlockwise).

Example-Rotate:

1) Construct drawing 1 of Figure 7.5.26 with Polyline. Copy the drawing 1 three times (Figure 7.5.26).

2) Call Rotate-left-click on its tool icon in the Home/Modify panel (Figure 7.5.27), pick its tool icon from the Modify toolbar, pick Rotate from the Modify drop-down menu, or enter ro or rotate at the command line. The command line shows:

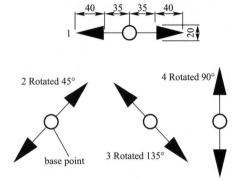

Figure 7.5.26　Example——Rotate

Figure 7.5.27　The Rotate tool icon from the Home/Modify panel

Command: _ rotate.

Current positive angle in UCS: ANGDIR= counterclockwise ANGBASE=0.

Select objects: window the drawing 3 found

Select objects: right-click.

Specify base point: pick.

Specify rotation angle or [Copy/Reference] <0>: 45.

And the first copy rotates through the specified angle.

Repeat for drawings 3 and 4 rotating as shown in Figure 7.5.26.

8. The Scale tool

Examples——Scale:

1) Using the Rectangle and Polyline tools, construct drawing 1 of Figure 7.5.28. The Rec-

tangle fillets are R10. The line width of all parts is 1. Copy the drawing 3 times to give drawings 2, 3 and 4.

2) Call Scale——left-click on its tool icon in the Home/Draw panel (Figure 7.5.29),

pick its tool icon in the Modify toolbar, pick Scale from the Modify drop-down menu, or click its tool in the Modify toolbar, or enter sc or scale at the command line which then shows:

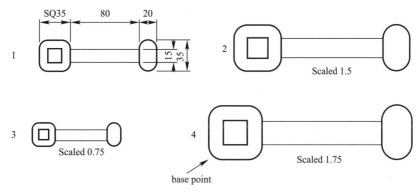

Figure 7.5.28　Examples——Scale

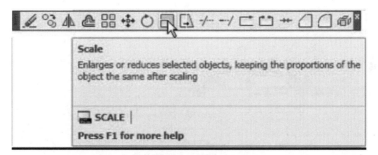

Figure 7.5.29　The Scale tool icon from the Modify toolbar

Command: _ scale.

Select objects: window drawing 2 5 found Select objects: right-click.

Specify base point: pick.

Specify scale factor or [Copy/Reference] <1>: 0.75.

3) Repeat for the other two drawings 3 and 4 scaling to the scales given with the drawings.

9. The Trim tool

This tool is one which will be in frequent use when constructing drawings.

1) First example——Trim

(1) Construct the drawing Original drawing in Figure 7.5.30.

(2) Call Trim——either left-click on its tool icon in the Home/Modify panel (Figure 7.5.31), pick its tool icon in the Modify toolbar, pick Trim from the Modify drop-down menu, or enter tr or trim at the command line, which then shows:

Command: _ trim.

Current settings: Projection UCS. Edge = Extend Select cutting edges.

Select objects or < select all >: pick the left-hand circle 1 found.

Select objects to trim or shift-select to extend or [Fence/Project/Crossing/Edge/Erase/Undo]: pick one of the objects.

Select objects to trim or shift-select to extend or [Fence/Crossing/Project/Edge/Erase/Undo]:. pick the second of the objects.

Select objects to trim or shift-select to extend or [Project/Edge/Undo]: right-click.

(3) This completes the First stage as shown in Figure 7.5.30. Repeat the Trim sequence for the Second stage.

(4) The Third stage drawing of Figure 7.5.30 shows the result of the trims at the left-hand end of the drawing.

(5) Repeat for the right-hand end. The final result is shown in the drawing labelled Result in Figure 7.5.30.

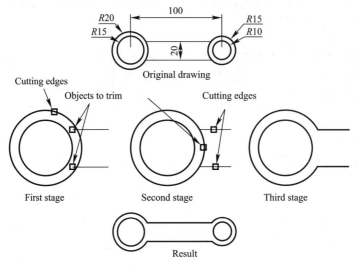

Figure 7.5.30　First example——Trim

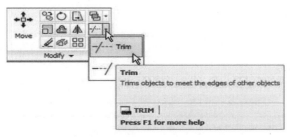

Figure 7.5.31　The Trim tool icon from the Modify toolbar in the Home/Modify panel

2) Second example——Trim

(1) Construct the left-hand drawing of Figure 7.5.32.

(2) Call Trim. The command line shows:
Command: _ trim.

Current settings: Projection UCS. Edge = Extend Select cutting edges.

Select objects or <select all>: pick the left-hand arc 1 found.

Select objects: right-click.

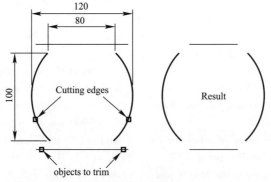

Figure 7.5.32　Second example——Trim

Select objects to trim or shift-select to extend or [Fence/Crossing/Project/Edge/Erase/Undo]: e (Edge).

Enter an implied edge extension mode [Extend/No extend] <No extend>: e (Extend).

Select objects to trim: pick Select objects to trim: pick.

Select objects to trim: right-click.

(3) Repeat for the other required trims. The result is given in Figure 7.5.32.

10. The Stretch tool

Examples——Stretch:

As its name implies the Stretch tool is for stretching drawings or parts of drawings. The action of the tool prevents it from altering the shape of circles in any way. Only crossing or polygonal windows can be used to determine the part of a drawing which is to be stretched.

1) Construct the drawing labelled Original in Figure 7.5.33, but do not include the dimensions. Use the Circle, Arc, Trim and Edit Polyline tools. The resulting outlines are plines of width=1. With the Copy tool make two copies of the drawing.

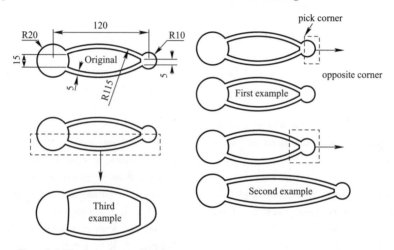

Figure 7.5.33 Examples——Stretch

Note:

In each of the three examples in Figure 7.5.33, the broken lines represent the crossing windows required when Stretch is used.

2) Call the Stretch tool——either click on its tool icon in the Home/Modify panel (Figure 7.5.34), left-click on its tool icon in the Modify toolbar, pick its name in the Modify drop-down menu, or enter s or stretch at the command line, which shows:

Command: _ stretch.

Select objects to stretch by crossing-window or crossing-polygon.

Select objects: enter c right-click.

Specify first corner: pick Specify opposite corner: pick 1 found.

Select objects: right-click.

Specify base point or [Displacement] <Displacement>: pick beginning of arrow Specify second point of displacement or <use first point as displacement>: drag in the direction of the arrow to the required second point and right-click.

Notes:

(1) When circles are windowed with the crossing window no stretching can take place. This is why, in the case of the first example in Figure 7.5.33, when the second point of displacement was picked, there was no result——the outline did not stretch.

(2) Care must be taken when using this tool as unwanted stretching can occur.

11. The Break tool

Examples——Break:

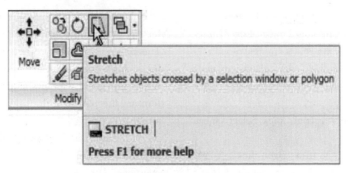

Figure 7.5.34 The Stretch tool icon from the Home/Modify panel

1) Construct the rectangle, arc and circle (Figure 7.5.35).

2) Call Break——either click on its tool icon in the Home/Modify panel (Figure 7.5.36), pick its tool icon in the Modify toolbar, click Break in the Modify drop-down menu, or enter br or break at the command line, which shows:

For drawings 1 and 2:

Command: _ break Select object: pick at the point Specify second break point or [First point]: pick.

For drawing 3:

Command: _ break Select object pick at the point Specify second break point or [First point]: enter f right-click.

Specify first break point: pick Specify second break point: pick.

The results are shown in Figure 7.5.35.

Note:

Remember the default rotation of AutoCAD is counterclockwise. This applies to the use of the Break tool.

12. The Join tool

The Join tool can be used to join plines providing their ends are touching; to join lines which are in line with each other; to join arcs and convert arcs to circles.

Examples——Join:

1) Construct a rectangle from four separate plines——drawing 1 of Figure 7.5.37. Construct two lines——drawing 2 of Figure 7.5.37 and an arc——drawing 3 of Figure 7.5.37.

2) Call the Join tool——either click on the Join tool icon in the Home/ Modify panel (Figure 7.5.38), left-click its tool icon in the Modify toolbar, select Join from the Modify drop-down menu, or enter join or j at the com

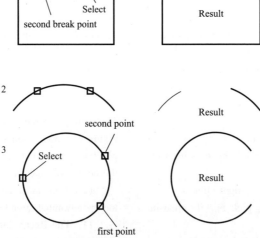

Figure 7.5.35 Examples——Break

mand line. The command line shows:

Figure 7.5.36 The Break tool icon from the Home/Modify pane

Command: _ join Select source object.

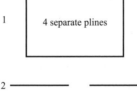

Figure 7.5.37 Examples——Join

Join Select source object: pick the arc.

Select arcs to join to source or [Close]: enter right-click.

Figure 7.5.38 The Join tool icon from the Home/Modify panel

Arc converted to a circle.

The results are shown in Figure 7.5.37.

13. The Extend tool

Examples——Extend:

Select objects to join to source: pick a pline 1 found.

Select objects to join to source: pick another 1 found, 2 total.

Select objects to join to source: pick another 1 found, 3 total.

Select objects to join to source: right-click.

3) Segments added to polyline:

Command: right-click.

Join Select source object: pick one of the lines Select lines to join to source: pick the other 1 found Select lines to join to source: right-click.

1 line joined to source Command: right-click.

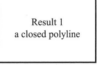

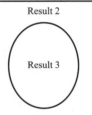

1) Construct plines and a circle as shown in the left-hand drawings of Figure 7.5.39.

2) Call Extend——either click on the Extend tool icon in the Modify toolbar (Figure 7.5.40), pick Extend from the Modify drop-down menu, or enter ex or extend at the command line which then shows:

Command: _ extend.

Current settings: Projection = UCS Edge = Extend Select boundary edges.

Select objects or <select all>: pick 1 found Select objects: right-click.

Select object to extend or shift-select to trim or [Fence/Crossing/Project/Edge/Undo]: pick.

Repeat for each object to be extended.

Select object to extend or shift-select to trim or [Fence/Crossing/Project/Edge/Undo]: right-click.

The results are shown in Figure 7.5.39.
Note:

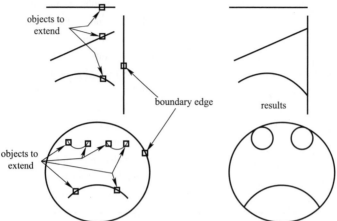

Figure 7.5.39 Examples——Extend

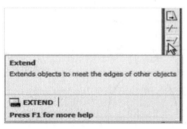

Figure 7.5.40 The Extend tool icon in the Modify toolbar in the AutoCAD Classic workspace

14. The Fillet and Chamfer tools

These two tools can be called from the Home/Modify panel. There are similarities in the prompt sequences for these two tools. The major differences are that only one (Radius) setting is required for a fillet, but two (Dist1 and Dist2) are required for a chamfer. The basic prompts for both are:

1) Fillet

Command: _ fillet.

Current settings: Mode = TRIM, Radius = 1 Select first object or [Polyline/Radius/Trim/Multiple]: enter r (Radius) right-click.

Specify fillet radius <1>: 15.

2) Chamfer

Command: _ chamfer.

(TRIM mode) Current chamfer Dist1 = 1, Dist2 = 1 Select first line or [Undo/Polyline/Distance/Angle/Trim/Method/Multiple]: enter d (Distance) right-click.

Specify first chamfer distance <1>: 10 Specify second chamfer distance <10>: right-click.

Examples——Fillet:

(1) Construct three rectangles 100 by 60 using either the Line or the Polyline tool (Figure 7.5.41).

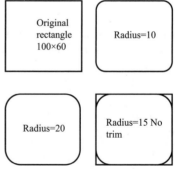

Figure 7.5.41 Examples——Fillet

(2) Call Fillet——click the arrow to the right of the tool icon in the Home/ Modify panel and select Fillet from the menu which appears (Figure 7.5.42), pick its tool icon in the Modify toolbar, pick Fillet from the Modify drop-down menu, or enter f or fillet at the command line which then shows:

Command: _ fillet.

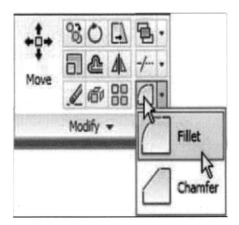

Figure 7.5.42 Select Fillet from the menu in the Home/Modify panel

Current settings: Mode = TRIM, Radius = 1 Select first object or [Polyline/Radius/Trim/Multiple]: r (Radius) Specify fillet radius <0>: 15.

Select first object or [Undo/Polyline/Radius/Trim/ Multiple]: pick.

Select second object or shift-select to apply corner: pick.

Three examples are given in Figure 7.5.41.

Examples——Chamfer:

(1) Construct three rectangles 100 by 60 using either the Line or the Polyline tool.

(2) Call Chamfer——click the arrow to the right of the tool icon in the Home/Modify panel and select Chamfer from the menu which appears (Figure 7.5.43), click on its tool icon in the Modify toolbar, pick Chamfer from the Modify drop-down menu, or enter cha or chamfer at the command line which then shows:

Command: _ chamfer.

(TRIM mode) Current chamfer Dist1 = 1, Dist2 = 1 Select first line or [Undo/Polyline/Distance/Angle/Trim/Method/Multiple]: d.

Specify first chamfer distance <1>: 10 Specify second chamfer distance <10>: right-click.

Select first line or [Undo/Polyline/Distance/Angle/Trim/Method/Multiple]: pick the first line for the chamfer.

Select second line or shift-select to apply corner: pick.

The result is shown in Figure 7.5.44. The other two rectangles are chamfered in a similar manner except that the No trim prompt is brought into operation with the bottom left-hand example.

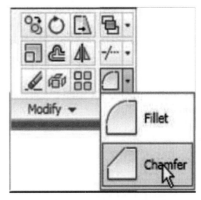

Figure 7.5.43 Select Chamfer from the Home/Modify panel

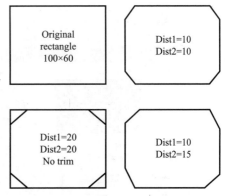

Figure 7.5.44 Examples——Chamfer

15. Revison notes

1) The Modify tools are among the most frequently used tools in AutoCAD.

2) The abbreviations for the Modify tools are:

Copy——cp or co;
Mirror——mi;
Offset——o;
Array——ar;
Move——m;
Rotate——ro;
Scale——sc;
Stretch——s;

Trim——**tr**;
Extend——**ex**;
Break——**br**;
Join——**j**;
Chamfer——**cha**;
Fillet——**f**.

3) There are two other tools in the Modify toolbar or in the 2D Draw control panel——Erase——some examples were given in Chapter 2 and Explode——further details of this tools will be given in Chapter 9.

A note——**selection windows and crossing windows**:

In the Options dialog settings can be made in the Selection sub-dialog for Visual Effects. A click on the Visual Effects Settings"……" button brings up another dialog. If the Area Selection Effect settings are set on a normal window from top left to bottom right, the window will colour in a chosen colour (default blue). A crossing window, bottom left to top right, will be coloured red (default colour). Note also that highlighting ——selection Preview Effect—— allows objects to highlight if this feature is on. These settings are shown in Figure 7.5.45.

4) When using Mirror, if text is part of the area to be mirrored, the set variable Mirrtext will require setting——to either 1 or 0.

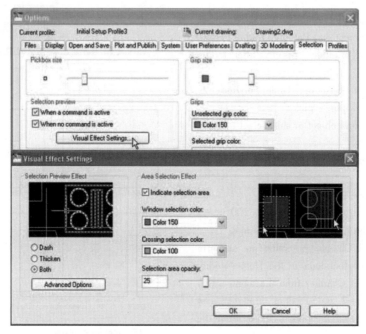

Figure 7.5.45 Visual Setting Effects Settings sub-dialog of the Options dialog

5) With Offset the Through prompt can be answered by clicking two points in the drawing area the distance of the desired offset distance.

6) Polar Arrays can be arrays around any angle set in the Angle of array field of the Array dialog.

7) When using Scale, it is advisable to practise the Reference prompt.

8) The Trim tool in either its Trim or its No trim modes is among the most useful tools in AutoCAD.

9) When using Stretch, circles are unaffected by the stretching.

10) There are some other tools in the Home/Modify panel not described in this book. The reader is invited to experiment with these other tools. They are:

Bring to Front, Send to Back, Bring above Objects, Send under Objects Set by Layer, Change Space, Lengthen, Edit Spline, Edit Hatch, Reverse.

7.6 Dimensions and Text

1. Introduction

The aims of this chapter are:

1) To describe a variety of methods of dimensioning drawings.

2) To describe methods of adding text to drawings.

The dimension style (My_ style) has already been set in the acadiso. dwt template, which means that dimensions can be added to drawings using this dimension style.

2. The Dimension tools

There are several ways in which the dimension tools can be called.

1) From the Annotate/Dimensions panel (Figure 7.6.1).

2) From the Dimension toolbar (Figure 7.6.2).

3) Click Dimension in the menu bar. Dimension tools can be selected from the drop-down menu which appears.

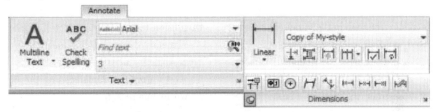

Figure 7.6.1 Dimension tools in the Annotate/Dimension panel

Figure 7.6.2 The Dimension toolbar

4) By entering an abbreviation for a dimension tool at the command line.

Any one of these methods can be used when dimensioning a drawing, but some operators may well decide to use a combination of the four methods.

3. Adding dimensions using the tools

1) First example——Linear Dimension

(1) Construct a rectangle 180×110 using the Polyline tool.

(2) Make the Dimensions layer current (Home/Layers panel).

(3) Click the Linear tool icon in the Annotate/Dimension panel (Figure 7.6.3) or on Linear in the Dimension toolbar. The command line shows:

Command: _ dimlinear.

Specify first extension line origin or <select object>: pick.

Specify second extension line origin: pick

Non-associative dimension created.

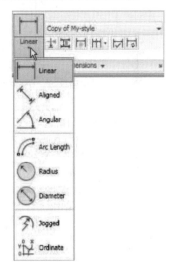

Figure 7.6.3 The Linear tool icon in the Annotate/Dimension panel

Specify dimension line location or [Mtext/Text/Angle/Horizontal/Vertical/Rotated]: pick.

Dimension text = 180.

Figure 7.6.4 shows the 180 dimension. Follow exactly the same procedure for the 110 dimension.

Notes:

(1) If necessary use Osnaps to locate the extension line locations.

(2) The prompt Specify first extension line origin or [select object]: also allows the line being dimensioned to be picked.

(3) The drop-down menu from the Line tool icon contains the following tool icons——Angular, Linear, Aligned, Arc Length, Radius, Diameter, Jog Line and Ordinate. Please refer to Figure 7.6.3 when working through the examples below. Note when a tool is chosen from this menu, the icon in the panel changes to the selected tool icon.

2) Second example——Aligned Dimension

(1) Construct the outline Figure 7.6.5 using the Line tool.

(2) Make the Dimensions layer current (Home/Layers panel).

(3) Left-click the Aligned tool icon and dimension the outline. The prompts and replies are similar to the first example.

3) Third example——Radius Dimension (Figure 7.6.6)

(1) Construct the outline using the Line and Fillet tools.

(2) Make the Dimensions layer current (Home/Layers panel).

(3) Left-click the Radius tool icon. The command line shows:

Command: _ dimradius.

Select arc or circle: pick one of the arcs Dimension text = 30.

Specify dimension line location or [Mtext/Text/Angle]: pick.

(4) Continue dimensioning the outline as shown in Figure 7.6.6.

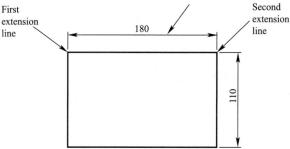

Figure 7.6.4　First example——Linear dimension

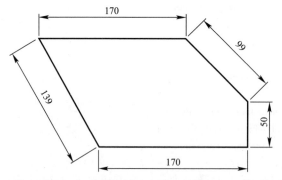

Figure 7.6.5　Second example——Aligned dimension

Notes:

(1) At the prompt: [Mtext/Text/Angle]: If a t (Text) is entered, another number can be entered, but remember if the dimension is a ra-

dius the letter R must be entered as a prefix to the new number.

(2) If the response is a (Angle), and an angle number is entered the text for the dimension will appear at an angle.

(3) If the response is m (Mtext) the Text Formatting dialog appears together with a box in which new text can be entered.

(4) Dimensions added to a drawing using other tools from the Dimensions control panel or from the Dimension toolbar should be practised.

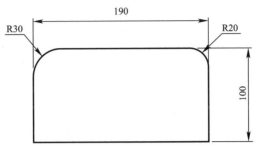

Figure 7.6.6 Third example——Radius dimension

4. Adding dimensions from the command line

From Figure 7.6.1 to Figure 7.6.2 it will be seen that there are some dimension tools which have not been described in examples. Some operators may prefer entering dimensions from the command line. This involves abbreviations for the required dimension such as:

For Linear Dimension——hor (horizontal) or ve (vertical);

For Aligned Dimension——al;

For Radius Dimension——ra;

For Diameter Dimension——d;

For Angular Dimension——an;

For Dimension Text Edit——te;

For Quick Leader——l.

And to exit from the dimension commands——e (Exit).

1) First example——hor and ve (horizontal and vertical)

(1) Construct the outline Figure 7.6.7 using the Line tool. Its dimensions are shown in Figure 7.6.8.

(2) Make the Dimensions layer current

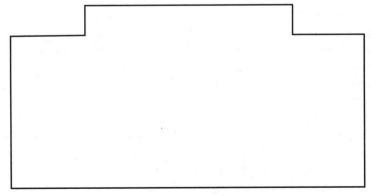

Figure 7.6.7 First example——outline to dimension

(Home/Layers panel).

(3) At the command line enter dim. The command line will show:

Command: enter dim right-click.

Dim: enter hor (horizontal) right-click.

Specify first extension line origin or <select object>: pick.

Specify second extension line origin: pick

Non-associative dimension created.

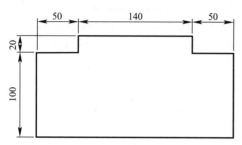

Figure 7.6.8 First example——horizontal and vertical dimensions

Specify dimension line location or [Mtext/Text/Angle]: pick.

Enter dimension text <50>: right-click.

Dim: right-click HORIZONTAL.

Specify first extension line origin or <select object>: pick.

Specify second extension line origin: pick Non-associative dimension created.

Specify dimension line location or [Mtext/Text/Angle/Horizontal/Vertical/Rotated]: pick Enter dimension text <140>: right-click.

Dim: right-click.

And the 50 and 140 horizontal dimensions are added to the outline.

(4) Continue to add the right-hand 50 dimension. Then when the command line shows:

Dim: enter ve (vertical) right-click.

Specify first extension line origin or <select object>: pick.

Specify second extension line origin: pick Specify dimension line location or [Mtext/Text/Angle/Horizontal/Vertical/Rotated]: pick Dimension text <20>: right-click.

Dim: right-click.

VERTICAL:

Specify first extension line origin or <select object>: pick.

Specify second extension line origin: pick Specify dimension line location or [Mtext/Text/Angle/Horizontal/Vertical/Rotated]: pick Dimension text <100>: right-click.

Dim: enter e (Exit) right-click.

The result is shown in Figure 7.6.8.

2) Second example——an (Angular)

(1) Construct the outline Figure 7.6.9——a pline of width = 1.

(2) Make the Dimensions layer current (Home/Layers panel).

(3) At the command line:

Command: enter dim right-click.

Dim: enter an right-click.

Select arc, circle, line or <specify vertex>: pick.

Select second line: pick.

Specify dimension arc line location or [Mtext/Text/Angle/Quadrant]: pick.

Enter dimension <90>: right-click.

Enter text location (or press ENTER): pick.

And so on to add the other angular dimensions. The result is given in Figure 7.6.10.

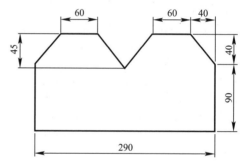

Figure 7.6.9 Second example——outline for dimensions

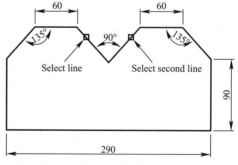

Figure 7.6.10 Second example——an (Angular) dimension

3) Third example-l (Leader)

(1) Construct Figure 7.6.11.

(2) Make the Dimensions layer current (Home/Layers panel).

(3) At the command line:

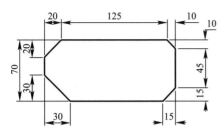

Figure 7.6.11 Third example——outline for dimensioning

Command: enter dim right-click.
Dim: enter l (Leader) right-click.
Leader start: enter nea (osnap nearest) right-click to pick one of the chamfer lines.
To point: pick.
To point: pick.
To point: right-click.
Dimension text <0>: enter CHA 10 × 10 right-click.
Dim: right-click.
Continue to add the other leader dimensions-Figure 7.6.12.

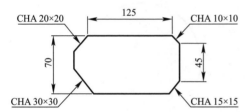

Figure 7.6.12 Third example——l (Leader) dimensions

4) Fourth example——te (Dimension Text Edit)
(1) Construct Figure 7.6.13.

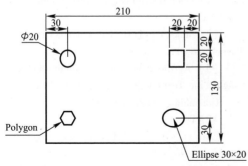

Figure 7.6.13 Fourth example——dimensioned drawing

(2) Make the Dimensions layer current (Home/Layers panel).

(3) At the command line:
Command: enter dim right-click.
Dim: enter te (tedit) right-click.
Select dimension: pick the dimension to be changed Specify new location for text or [Left/Right/Center/Home/Angle]: either pick or enter a prompt capital letter.
Dim: The results as given in Figure 7.6.14 show dimensions which have been moved. The 210 dimension changed to the left-hand end of the dimension line, the 130 dimension changed to the left-hand end of the dimension line and the 30 dimension position changed.

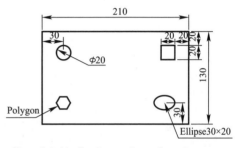

Figure 7.6.14 Fourth example——dimensions amended with te (Dimension Text Edit)

5. The Arc Length tool

1) Construct two arcs of different sizes as in Figure 7.6.15.

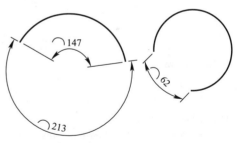

Figure 7.6.15 Examples——Arc Length tool

2) Make the Dimensions layer current (Home/Layers panel).

3) Call the Arc Length tool from the Annotate/Dimensions panel, or click on Arc Length in the Dimension toolbar, or enter dimarc at the command line.

The command line shows:
Command: _ dimarc.

Select arc or polyline arc segment: pick an arc Specify arc length dimension location, or [Mtext/Text/Angle/Partial/Leader]: pick a suitable position.

Dimension text = 147.

Examples on two arcs are shown in Figure 7.6.15.

6. The Jogged tool

1) Draw a circle and an arc as indicated in Figure 7.6.16.

2) Make the Dimensions layer current (Home/Layers panel).

3) Call the Jogged tool, either with a left-click on its tool icon in the Annotation/Dimension panel, or with a click on Jogged in the Dimension toolbar, or by entering jog at the command line. The command line shows:

Command: _ dim jogged.

Select arc or circle: pick the circle or the arc Specify center location override: pick.

Dimension text = 60.

Specify dimension line location or [Mtext/Text/Angle]: pick.

Specify jog location: pick.

The results of placing a jogged dimension on a circle and an arc are shown in Figure 7.6.16.

7. Dimension tolerances

Before simple tolerances can be included with dimensions, new settings will need to be made in the Dimension Style Manager dialog as follows:

1) Open the dialog. The quickest way of doing this is to enter d at the command line followed by a right-click. This opens up the dialog.

2) Click the Modify······button of the dialog, followed by a left-click on the Primary Units tab and in the resulting sub-dialog make settings as shown in Figure 7.6.17. Note the changes in the preview box of the dialog.

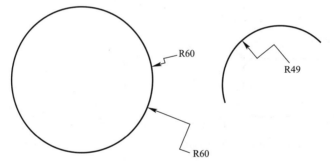

Figure 7.6.16　Examples——the Jogged tool

Example——tolerances:

1) Construct the outline Figure 7.6.18.

2) Make the Dimensions layer current (Home/Layers panel).

3) Dimension the drawing using either tools from the Dimension toolbar or by entering abbreviations at the command line. Because tolerances have been set in the Dimension Style Manager dialog (Figure 7.6.17), the toleranced dimensions will automatically be added to the drawing (Figure 7.6.19).

8. Text

There are two main methods of adding text to drawings——Multiline Text and Single Line Text.

1) First example——Single Line Text

(1) Open the drawing from the example on tolerances.

(2) Make the Text layer current (Home/Layers panel).

(3) At the command line enter dt (for Single Line Text) followed by a right-click.

Command: enter dt right-click.

TEXT

Current text style "ARIAL" Text height: 8 Annotative No.

Specify start point of text or [Justify/Style]: pick.

Specify rotation angle of text <0>: right-click.

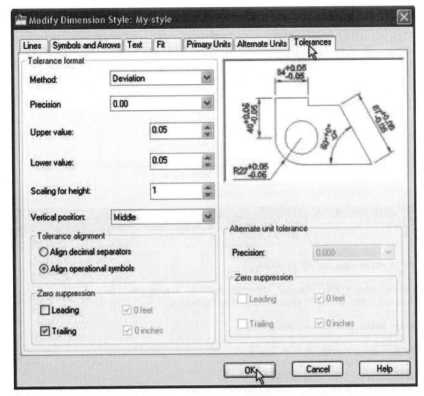

Figure 7.6.17 The Tolerances sub——dialog of the Modify Dimension Style dialog

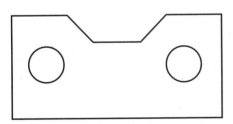

Figure 7.6.18 Example——simple tolerances - outline

Enter text: enter The dimensions in this drawing show tolerances press the Return key twice.

The result is given in Figure 7.6.20.

Notes:

(1) When using Dynamic Text the Return key of the keyboard is pressed when the text has been entered. A right-click does not work.

(2) At the prompt:

Specify start point of text or [Justify/Style]: enter s (Style) right-click.

Enter style name or [?] <ARIAL>: enter?

right -click.

Enter text style(s) to list < * >: right-click.

And an AutoCAD Text Window (Figure 7.6.20) appears listing all the styles which have been selected in the Text Style dialog.

(3) In order to select the required text style its name must be entered at the prompt:

Enter style name or [?] <ARIAL>: enter Romand right-click.

And the text entered will be in the Romand style of height 9, but only if that style was previously selected in the Text Style dialog.

(4) Figure 7.6.21 shows some text styles from the AutoCAD Text Window.

(5) There are two types of text fonts available in AutoCAD——the AutoCAD SHX fonts and the Windows True Type fonts. The ITALIC, ROMAND, ROMANS and STANDARD styles shown in Figure 7.6.21 are AutoCAD text fonts. The TIMES and ARIAL styles are Windows True Type styles. Most of the True

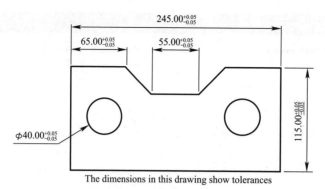

Figure 7.6.19 Example——tolerances

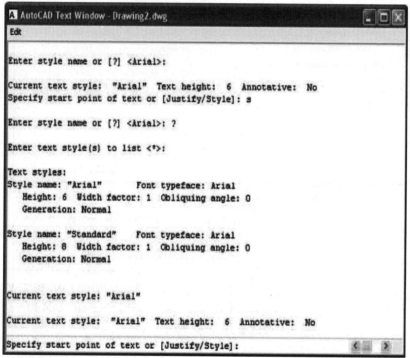

Figure 7.6.20 The AutoCAD Text Window

This is the TIMES text
This is ROMANC text
This is ROMAND text
This is STANDARD text
This is ITALIC text
This is ARIAL text

Figure 7.6.21 Some text styles

Type fonts can be entered in Bold, Bold Italic, Italic or Regular styles, but these variations are not possible with the AutoCAD fonts.

(6) In the Font name pop-up list of the Text Style dialog, it can be seen that a large number of text styles are available to the Auto-CAD operator. It is advisable to practise using a variety of these fonts to familiarise oneself with the text opportunities available with AutoCAD.

2) Second example——Multiline Text

(1) Make the Text layer current (Home/Layers panel).

(2) Either left-click on the Multiline Text tool icon in the Home/Annotation panel (Figure 7.6.22), or click on Multiline Text······in the Draw toolbar, or enter t at the command line:

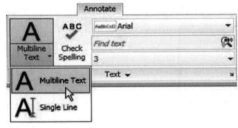

Figure 7.6.22 Selecting Multiline Text······from the Home/Annotation panel

Command: _ mtext.

Current text style: "Arial" Text height: 6 Annotative No.

Specify first corner: pick.

Specify opposite corner or [Height/Justify/Line spacing/Rotation/Style/Width/Columns]: pick.

As soon as the opposite corner is picked, the Text Formatting box appears (Figure 7.6.23). Text can now be entered as required within the box as indicated in this illustration.

When all the required text has been entered left-click and the text box disappears leaving the text on screen.

9. Symbols used in text

When text has to be added by entering letters and figures as part of a dimension, the following symbols must be used:

To obtain $\phi 75$ enter %%c75;

To obtain 55% enter 55%%%;

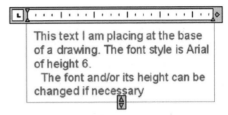

Figure 7.6.23 Example——Multiline Text entered in the text box

To obtain ±0.05 enter %%p 0.05;

To obtain 90° enter 90%%d.

10. Checking spelling

Note:

When a misspelt word or a word not in the AutoCAD spelling dictionary is entered in the Multiline Text box, red dots appear under the word, allowing immediate correction.

There are two methods for the checking of spelling in AutoCAD.

1) First example——spell checking-ddedit

(1) Enter some badly spelt text as indicated in Figure 7.6.24.

(2) Enter ddedit at the command line.

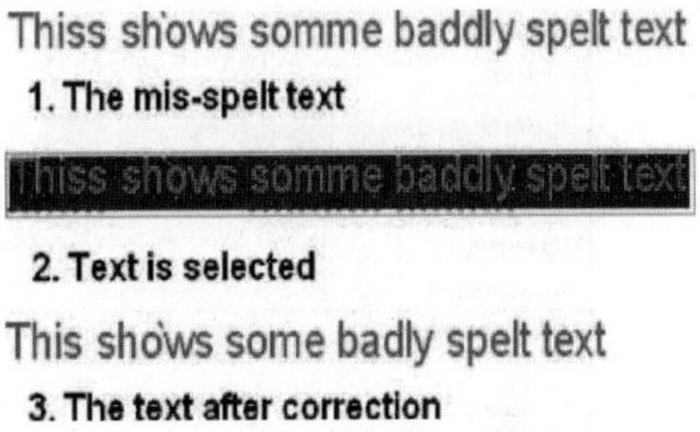

Thiss shows somme baddly spelt text
1. The mis-spelt text

Thiss shows somme baddly spelt text
2. Text is selected

This shows some badly spelt text
3. The text after correction

Figure 7.6.24 First example——spell checking——ddedit

(3) Left-click on the text. The text is highlighted. Edit the text as if working in a word-processing application and when satisfied left-click followed by a right-click.

2) Second example——the Spelling tool

(1) Enter some badly spelt text as indicated in Figure 7.6.25.

(2) Either click the Spell Check"···"icon in the Annotate/Text panel (Figure 7.6.26) or enter spell or sp at the command line.

(3) The Check Spelling dialog appears (Figure 7.6.26). In the Where to look field

select Entire drawing from the field's pop-up list. The first badly spelt word is highlighted with words to replace them listed in the Suggestions field. Select the appropriate correct spelling as shown. Continue until all text is checked. When completely checked an AutoCAD Message appears (Figure 7.6.27). If satisfied click its OK button.

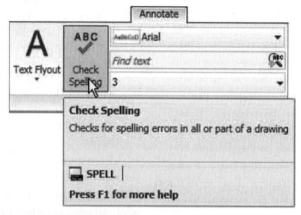

Figure 7.6.25　Second example——the Check Spelling dialog

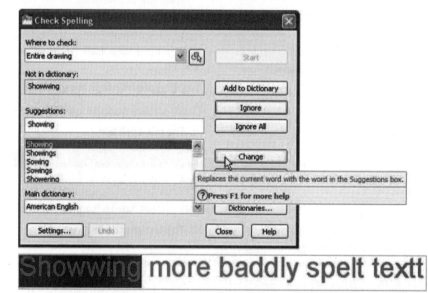

Figure 7.6.26　The Spell Check...... icon in the Annotate/Text penel

Figure 7.6.27　The AutoCAD Message window showing that spelling check is complete

11. Revison notes

1) In the Line and Arrows sub-dialog of the Dimension Style Manager dialog Lineweights were set to 0.3. If these lineweights are to show in the drawing area of AutoCAD, the Show/Hide Lineweight button in the status bar must be set ON.

2) Dimensions can be added to drawings using the tools from the Annotate/Dimensions

panel, from the Dimension toolbar, or by entering dim, followed by abbreviations for the tools at the command line.

3) It is usually advisable to use Osnaps when locating points on a drawing for dimensioning.

4) The Style and Angle of the text associated with dimensions can be changed during the dimensioning process.

5) When wishing to add tolerances to dimensions it will probably be necessary to make new settings in the Dimension Style Manager dialog.

6) There are two methods for adding text to a drawing——Single Line Text and Multiline Text.

7) When adding Single Line Text to a drawing, the Return key must be used and not the right-hand mouse button.

8) Text styles can be changed during the process of adding text to drawings.

9) AutoCAD uses two types of text style——AutoCAD SHX fonts and Windows True Type fonts.

10) Most True Type fonts can be in bold, bold italic, italic or regular format. AutoCAD fonts can only be added in the single format.

11) To obtain the symbols ϕ; ±; °; % use %%c; %%p; %%d; %%% before the figures of the dimension.

12) Text spelling can be checked by selecting Text/Edit...... from the Modify drop-down menu, by selecting Spell Check ...... from the Text control panel, or by entering spell or sp at the command line.

7.7 Hatching

1. Introduction

To give further examples of the use of hatching in its various forms.

In Section 7 an example of the hatching of a sectional view in an orthographic projection was given. Further examples of the use of hatching will be described in this chapter.

There are a large number of hatch patterns available when hatching drawings in AutoCAD. Some examples from the Other Predefined set of hatch patterns (Figure 7.7.1) in the Hatch Pattern Palette sub-dialog are shown in Figure 7.7.2.

Other hatch patterns can be selected form the ISO or ANSI hatch pattern palettes, or the operator can design his/her own hatch patterns and save them to the Custom hatch palette.

1) First Example——hatching a sectional view

Figure 7.7.3 shows a two-view orthographic projection which includes a sectional end view. Note the following in the drawing:

(1) The section plane line, consisting of a centre line with its ends marked A and arrows showing the direction of viewing to obtain the sectional view.

(2) The sectional view labelled with the letters of the section plane line.

(3) The cut surfaces of the sectional view hatched with the ANSI31 hatch pattern, which is in general use for the hatching of engineering drawing sections.

2) Second example——hatching rules

Figure 7.7.4 describes the stages in hatching a sectional end view of a lathe tool holder. Note the following in the section:

(1) There are two angles of hatching to differentiate separate parts of the section.

(2) The section follows the general rule that parts such as screws, bolts, nuts, rivets, other cylindrical objects, webs and ribs and other such features are shown as outside views within sections.

3) Third example——Associative hatching

Figure. 7.7.5 shows two end views of a house. After constructing the left-hand view, it was found that the upper window had been placed in the wrong position. Using the Move

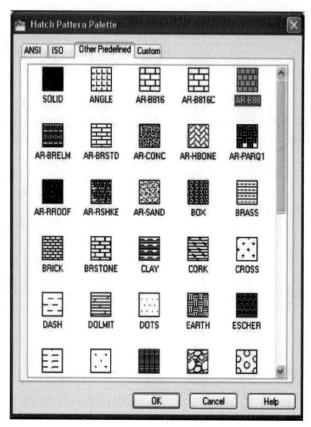

Figure 7.7.1 The Other Predefined Hatch Pattern Palette

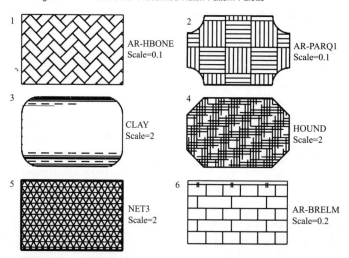

Figure 7.7.2 Some hatch patterns from Predefined hatch patterns

tool, the window was moved to a new position. The brick hatching automatically adjusted to the new position. Such associative hatching is only possible if check box is ON——a tick in the check box in the Options area of the dialog (Figure 7.7.6).

4) Fourth example——Colour Gradient hatching

Figure 7.7.8 shows two examples of hatching from the Gradient sub-dialog of the Hatch

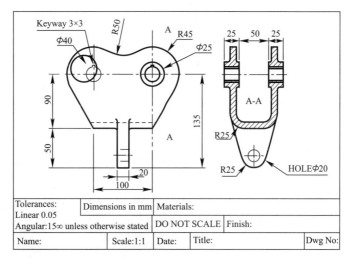

Figure 7.7.3 First example——Hatching

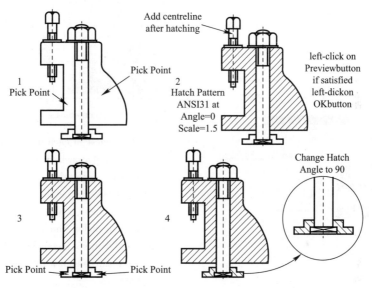

Figure 7.7.4 Second example——hatching rules for sections

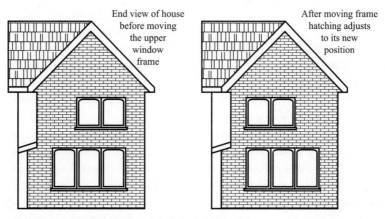

Figure 7.7.5 Third example——Associative hatching

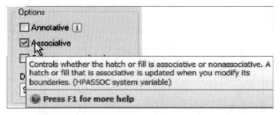

Figure 7.7.6   Associative hatching set ON in the Hatch and Gradient dialog

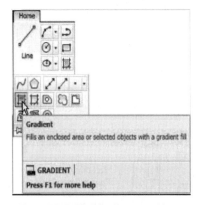

Figure 7.7.7   The Gradient…… tool icon

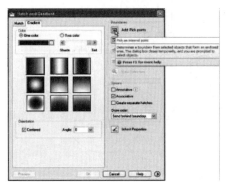

Figure 7.7.8   The Hatch and Gradient dialog

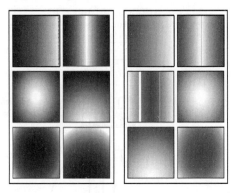

Figure 7.7.9   Fourth example——Gradient hatching

and Gradient dialog.

(1) Construct two outlines each consisting of six rectangles (Figure 7.7.9).

(2) Click the Gradient icon in the Home/Draw panel (Figure 7.7.7) or in the Draw toolbar. In the Hatch and Gradient dialog which appears from the Home/Draw panel (Figure 7.7.8) pick one of the gradient choices, followed with a click on the Pick an internal point button. Click one of the colour panels in the dialog and when then the dialog disappears, pick a single area of one of the rectangles in the left-hand drawings, followed by a click on the dialog's OK button when the dialog reappears.

(3) Repeat in each of the other rectangles of the left-hand drawing, changing the pattern in each of the rectangles.

(4) Click the button (……) to the right of the Colour field, select a new colour from the Select Colour dialog which appears and repeat items 3 and 4 in six rectangles.

The result is shown in Figure 7.7.9.

Note:

If the Two color radio button is set on (dot in circle) the colours involved in the gradient hatch can be changed by clicking the button marked with three full stops (……) on the right of the colour field. This brings a Select Color dialog on screen, which offers three choices of sub-dialogs from which to select colours.

5) Fifth example——advanced hatching

If the arrow at the bottom right-hand corner of the Hatch and Gradient dialog is clicked (Figure 7.7.10) the dialog expands to show the Island display style selections (Figure 7.7.11).

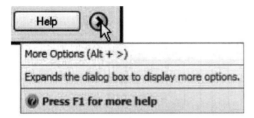

Figure 7.7.10 The More Options arrow of the Hatch and Gradient dialog

Figure 7.7.11 The Island display style selections in the expanded Hatch and Gradient dialog

(1) Construct a drawing which includes three outlines as shown in the left- hand drawing of Figure 7.7.12 and copy it twice to produce three identical drawings.

(2) Select the hatch patterns STARS at an angle of 0 and scale 1.

(3) Click in the Normal radio button of the Island display style area.

(4) Pick a point in the left-hand drawing. The drawing hatches as shown.

(5) Repeat in the centre drawing with the radio button of the Outer style set on (dot in button).

(6) Repeat in the right-hand drawing with Ignore set on.

6) Sixth example——text in hatching

(1) Construct a pline rectangle using the sizes given in Figure 7.7.13.

(2) In the Text Style Manager dialog, set the text font to Arial and its Height = 25.

(3) Using the Dtext tool enter the text as shown central to the rectangle.

(4) Hatch the area using the HONEY hatch pattern set to an angle of 0 and scale of 1.

The result is shown in Figure 7.7.13.

Note:

Text will be entered with a surrounding boundary area free from hatching providing the Normal radio button of the Island display style selection is on.

7) Seventh example——advanced hatching

(1) From the Home/Layers panel open the Layer list with a click on the arrow to the right of the Layer Control field (Figure 7.7.14).

(2) Note the extra added layer HATCH colour red (Figure 7.7.14).

(3) With the layer 0 current construct the outline as given in Figure 7.7.15.

(4) Make layer Text current and construct the lines as shown in Figure 7.7.16.

(5) Make the layer HATCH current and add hatching to the areas shown in Figure 7.7.17 using the hatch patterns ANGLE at scale 2 for the roof and BRICK at a scale of 0.75 for the wall.

(6) Finally turn the layer Text off. The result is given in Figure 7.7.18

2. REVISION NOTES

1) A large variety of hatch patterns are available when working with AutoCAD.

2) In sectional views in engineering drawings it is usual to show items such as bolts, screws, other cylindrical objects, webs and ribs as outside views.

3) When Associative hatching is set on, if an object is moved within a hatched area, the hatching accommodates to fit around the moved object.

4) Colour gradient hatching is available in AutoCAD.

5) When hatching takes place around text, a space around the text will be free from hatching.

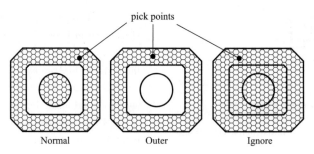

Figure 7.7.12　Fifth example——advanced hatching

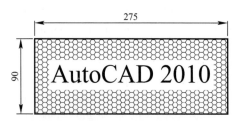

Figure 7.7.13　Sixth example——text in hatching

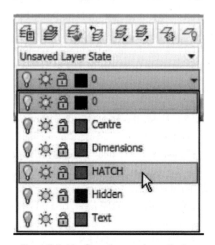

Figure 7.7.14　Seventh example——the layers setup for the advanced hatch example

Figure 7.7.15　Seventh example——construction on layer 0

Figure 7.7.16　Seventh example——construction on layer Text

Figure 7.7.17　Seventh example——construction on layer HATCH

Figure 7.7.18　Seventh example——the finished drawing

7.8 Building Drawings

1. Building drawings

The aims of this chapter are:

1) To show that AutoCAD is a suitable computer-aided design software package for the construction of building drawings.

2) To show that AutoCAD is a suitable CAD programme for the construction of 3D models of buildings.

There are a number of different types of drawings related to the construction of any form of building. In this chapter a fairly typical example of a set of building drawings is shown. These are seven drawings related to the construction of an extension to an existing two-storey house (44 Ridgeway Road). These show:

1) A site plan of the original two-storey house, drawn to a scale of 1 : 200 (Figure 7.8.1).

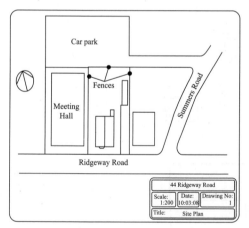

Figure 7.8.1   A site plan

2) A site layout plan of the original house, drawn to a scale of 1 : 100 (Figure 7.8.2).

3) Floor layouts of the original house, drawn to a scale of 1 : 50 (Figure 7.8.3).

4) Views of all four sides of the original house drawn to a scale of 1 : 50 (Figure 7.8.4).

5) Floor layouts including the proposed extension, drawn to a scale of 1 : 50 (Figure 7.8.5).

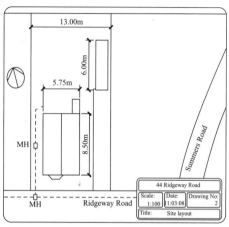

Figure 7.8.2   A site layout plan

6) Views of all four sides of the house including the proposed extension, drawn to a scale of 1 : 50 (Figure 7.8.6).

7) A sectional view through the proposed extension, drawn to a scale of 1 : 50 (Figure 7.8.7).

Notes:

1. Other types of drawings will be constructed such as drawings showing the details of parts such as doors, windows, floor structures etc. These are often shown in sectional views.

2. Although the seven drawings related to the proposed extension of the house at 44 Ridgeway Road are shown here as having been constructed on either A3 or A4 layouts, it is common practice to include several types of building drawings on larger sheets such as A1 sheets of a size 820 mm by 594 mm.

2. Floor layouts

When constructing floor layout drawings, it is advisable to build up a library of block drawings of symbols representing features such as doors, windows etc. These can then be inserted into layouts from the Design Center. A suggested small library of such block symbols in shown in Figure 7.8.8.

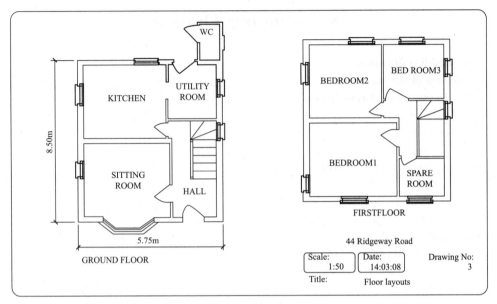

Figure 7.8.3 Floor layouts drawing of the original house

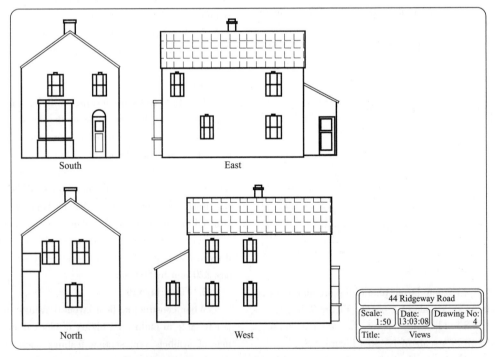

Figure 7.8.4 Views of the original house

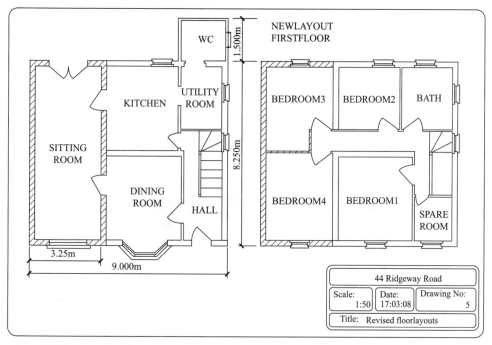

Figure 7.8.5 Floor layouts drawing of the proposed extension

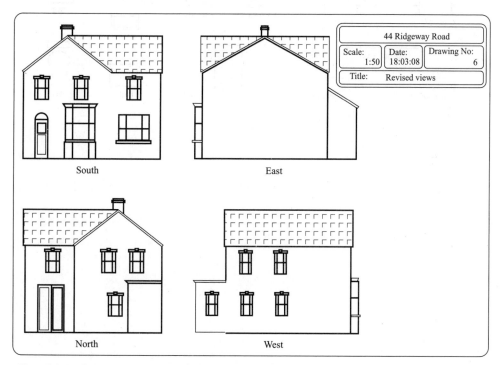

Figure 7.8.6 Views including the proposed extension

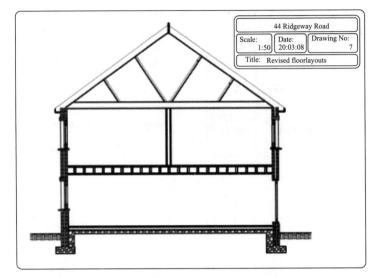

Figure 7.8.7　A section through the proposed extension

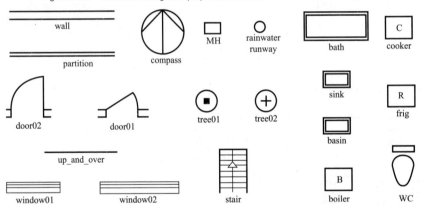

Figure 7.8.8　A small library of building symbols

References

[1] China Institute of Building Standard Design and Research. Unified Standard for Building Drawings (GB/T 50001—2017) [S]. Beijing: China Architecture& Building Press, 2018.

[2] China Institute of Building Standard Design and Research. Standard for General Layout Drawings (GB/T 50103—2010) [S]. Beijing: China Planning Press, 2011.

[3] China Institute of Building Standard Design and Research. Standard for Architectural Drawings (GB/T 50104—2010) [S]. Beijing: China Planning Press, 2011.

[4] China Institute of Building Standard Design and Research. Standard for Structural Drawings (GB/T 50105—2010) [S]. Beijing: China Architecture& Building Press, 2018.

[5] Highway Planning and Design Institute of the Ministry of Communications. Standard for Road Engineering Drawing (GB 50162—92) [S]. Beijing: China Planning Press, 1993.

[6] China Institute of Building Standard Design and Research. Drawing Rules and Standard Detailing Drawings of Ichnographic Representing Method for Construction Drawings of R. C. Structures (Cast-in-situ R. C. Frames, Shear Walls, Beams and Slabs) (16G101-1) [S]. Beijing: China Planning Press, 2017.

[7] China Institute of Building Standard Design and Research. Drawing Rules and Standard Detailing Drawings of Ichnographic Representing Method for Construction Drawings of R. C. Structures (Cast-in-situ Concrete Slab-stairs) (16G101-2) [S]. Beijing: China Planning Press, 2017.

[8] China Institute of Building Standard Design and Research. Drawing Rules and Standard Detailing Drawings of Ichnographic Representing Method for Construction Drawings of R. C. Structures (Spread Footings, Strip Foundations, Raft Foundation and Pile Foundations) (16G101-3) [S]. Beijing: China Planning Press, 2017.

[9] Yarwood A. Introduction to AutoCAD 2011 [M]. Amsterdam: Elsevier Ltd, 2010.

[10] Lu C X. Civil Engineering Drawing [M]. 5th edition. Beijing: China Architecture & Building Press, 2017.